Lecture Notes on

COMPLEX
ANALYSIS

Lecture Notes on

COMPLEX ANALYSIS

Ivan Francis Wilde

Imperial College Press

Published by

Imperial College Press
57 Shelton Street
Covent Garden
London WC2H 9HE

Distributed by

World Scientific Publishing Co. Pte. Ltd.
5 Toh Tuck Link, Singapore 596224
USA office: 27 Warren Street, Suite 401-402, Hackensack, NJ 07601
UK office: 57 Shelton Street, Covent Garden, London WC2H 9HE

British Library Cataloguing-in-Publication Data
A catalogue record for this book is available from the British Library.

LECTURE NOTES ON COMPLEX ANALYSIS

ISBN-13 978-1-86094-642-4
ISBN-10 1-86094-642-9
ISBN-13 978-1-86094-643-1 (pbk)
ISBN-10 1-86094-643-7 (pbk)

Printed in Singapore

To Erica

Preface

This text forms what is often referred to as "a first course in complex
analysis". It is a slight enhancement of lecture notes first presented to undergraduate students in the Mathematics Department of Bedford College,
University of London, as part of the Mathematics BSc. degree, and then
given for many years in the Mathematics Department of King's College,
London. During this time they have been continually revised, reorganized
and rewritten. The aim was to provide a rigorous and largely self-contained
but extremely gentle introduction to the basics of complex analysis.

The audience for the course comprised not only single subject mathematics BSc. and MSci. students but also a number of final year joint honours
students as well as postgraduate students who missed out on the subject
in their undergraduate programme.

There are a number of core topics (such as Cauchy's theorem, the Taylor
and Laurent series, singularities and the residue theorem) which simply
must be offered to any student of complex analysis. However, quite a
bit of preparation is required, so these important results unavoidably tend
to appear in rather rapid succession towards the end the course. This
leaves very little room for extra topics, especially if they are particularly
complicated or involve a lot of additional machinery. The presentation
here is for the benefit of the student audience. There has been no quest for
ultimate generality nor economy of delivery.

Nowadays, it seems that many students do not get to see an account
of metric spaces, so this aspect of complex analysis has been presented
in quite some detail (in Chapter 3). It is then but a small step for the
student wishing to go on to study metric spaces in general. The exponential
and trigonometric functions are defined via their power series expansions
in Chapter 5, so a certain amount of manœuvring is required to extract

those properties familiar from calculus—for example, the appearance of the number π is carefully explained. Those for whom this is familiar territory can quickly press on.

Most of the core results are contained in Chapters 8–12. The next two chapters, covering the maximum modulus principle and Möbius transformations have been moved around a bit over the years. For example, the maximum modulus principle (Chapter 13) could be discussed anytime after having dealt with Cauchy's integral formulae (Chapter 8). The treatment of Möbius transformations (Chapter 14) is essentially a stand-alone topic so could fit in almost anywhere. It might well be read after Chapter 8 so as to provide a little variety before embarking on the study of the Laurent expansion in Chapter 9.

It has to be admitted that the final section of Chapter 13 (on Hadamard's Theorem), possibly Chapter 15 (on harmonic functions) and Chapter 16 (on local properties of analytic functions) could be considered a bit of a luxury. In practice, they were all usually squeezed out because of lack of time. Most of the rest of the material in these notes just about fits into a one semester course.

The majority of students embarking on this subject will have studied calculus and will usually have also been exposed to some real analysis. Nevertheless, experience has shown that the odd reminder does not go amiss and so an appendix containing some pertinent facts from real analysis has been included. These are all consequences of the completeness property of $\mathbb{R}$ (so tend not to be very carefully covered in calculus courses—or else are deemed obvious).

A number of text books were consulted during the preparation of these notes and these are listed in the bibliography. No claim is made here regarding originality. As an undergraduate student in the Mathematics Department of Imperial College, it was my privilege to be taught analysis by M. C. Austin and Professor Ch. Pommerenke. Their lectures could only be described as both a joy and an inspiration. It is a pleasure to acknowledge my indebtedness to them both.

I. F. Wilde

Contents

Chapter 1

Complex Numbers

1.1 Informal Introduction

What is a complex number? It is any number of the form $z = x + iy$, where x and y are real numbers and i obeys $i^2 = -1$. Of course, there is no real number whose square is negative, and so i is not a real number. Accordingly, x is called the real part of z, denoted $\mathrm{Re}\,z$, and $y = \mathrm{Im}\,z$ is called the imaginary part. (Notice that $\mathrm{Im}\,z$ is y and not iy.)

Complex numbers are declared equal if and only if they have the same real and imaginary parts; if $z_1 = x_1 + iy_1$ and $z_2 = x_2 + iy_2$, then $z_1 = z_2$ if and only if both $x_1 = x_2$ and $y_1 = y_2$. We write 0 for $0 + i0$.

Addition and multiplication are as one would expect;

$$z_1 + z_2 = (x_1 + x_2) + i(y_1 + y_2)$$
$$z_1 z_2 = (x_1 + iy_1)(x_2 + iy_2) = x_1 x_2 + iy_1 x_2 + iy_1 iy_2 + x_1 iy_2$$
$$= (x_1 x_2 - y_1 y_2) + i(y_1 x_2 + x_1 y_2).$$

If $z = x + iy$, then $-z = -x - iy$.

Suppose that $z = x + iy$, and $z \neq 0$. Then at least one of x or y is non-zero. In fact, $z \neq 0$ if and only if $x^2 + y^2 > 0$. We have

$$\frac{1}{z} = \frac{1}{x + iy} = \frac{x - iy}{(x + iy)(x - iy)} = \frac{x - iy}{x^2 + y^2}$$
$$= \frac{x}{x^2 + y^2} - i\frac{y}{x^2 + y^2},$$

so that $\mathrm{Re}\,\dfrac{1}{z} = \dfrac{x}{x^2 + y^2}$ and $\mathrm{Im}\,\dfrac{1}{z} = \dfrac{-y}{x^2 + y^2}$.

By definition, complex conjugation changes the sign of the imaginary part, that is, the complex conjugate of $z = x + iy$ is defined to be the complex number $\overline{z} = x - iy$. Notice that $z\,\overline{z} = x^2 + y^2$ and that

$$\operatorname{Re} z = x = \frac{z + \overline{z}}{2} \quad \text{and} \quad \operatorname{Im} z = y = \frac{z - \overline{z}}{2i}.$$

Proposition 1.1   *For any complex numbers z_1, z_2, we have*

    (i) $\overline{z_1 + z_2} = \overline{z_1} + \overline{z_2},$

    (ii) $\overline{z_1 \, z_2} = \overline{z_1}\,\overline{z_2},$

    (iii) $\overline{\overline{z_1}} = z_1.$

    (iv)   *If $z_2 \neq 0$, then* $\overline{\left(\dfrac{z_1}{z_2}\right)} = \dfrac{\overline{z_1}}{\overline{z_2}}$.

Proof.   This is just straightforward computation. $\square$

1.2   Complex Plane

There is a natural correspondence between complex numbers and points in the plane, as follows. To any given complex number $z = x + iy$, we associate the point (x, y) in the plane and, conversely, to any point (x, y) in the plane, we associate the complex number $z = x + iy$

$$z = x + iy \quad \longleftrightarrow \quad (x, y).$$

This is evidently a one-one correspondence.

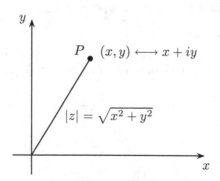

Fig. 1.1   Complex numbers as points in the Argand diagram.

The complex plane (also called the Argand diagram or Gauss plane) is just the set of complex numbers thought of as points in the plane in this way. It is very helpful to be able to picture complex numbers like this. The x-axis is called the real axis and the y-axis is called the imaginary axis.

If P is the point (x, y), corresponding to $z = x + iy$, then the (Euclidean) distance of P from the origin is equal to $\sqrt{x^2 + y^2}$. This value is written $|z|$, the modulus (or absolute value) of z. Thus, $|z|$ is the length of the two-dimensional vector (x, y). If z is real, then $y = 0$ and so $|z| = \sqrt{x^2} = |x|$, the usual value of the modulus of a real number.

For any complex numbers z_1 and z_2, $z_1 - z_2 = (x_1 - x_2) + i(y_1 - y_2)$ so that $|z_1 - z_2| = \sqrt{(x_1 - x_2)^2 + (y_1 - y_2)^2}$ which is the distance between the points z_1 and z_2 thought of as points in the plane. It makes perfectly good sense to talk about complex numbers being "close together"—this simply means that the distance between them, namely $|z_1 - z_2|$, is "small".

Examples 1.1

(1) What is the set $S = \{\, z : |z - \zeta| = r \,\}$, where $r > 0$ and ζ is fixed? The complex number z belongs this set if (and only if) its distance from ζ is equal to r. We conclude that S is the circle in the complex plane with centre ζ and radius r. In terms of cartesian coordinates, we see that $z = x + iy$ belongs to S if and only if

$$r^2 = |z - \zeta|^2 = (x - \xi)^2 + (y - \eta)^2$$

where $\zeta = \xi + i\eta$. This is the equation of a circle in $\mathbb{R}^2$ with centre at the point (ξ, η) and radius r.

By considering values $r < R$, we see that $\{\, z : |z - \zeta| < R \,\}$ is the disc in the complex plane formed by all those complex numbers whose distance from ζ is strictly less than R.

Note that $\{\, z : |z| = 1 \,\}$ is the circle with radius 1 and centre at the origin. The set $\{\, z : |z| < 1 \,\}$ is the disc with centre at the origin and radius 1 but *not* including the perimeter $\{\, z : |z| = 1 \,\}$.

(2) What is the set $A = \{\, z : |z - i| = |z - 3| \,\}$? We see that $z \in A$ if and only if its distance from the complex number i is the same as its distance from 3. It follows that A is a straight line—the perpendicular bisector of the line between i and 3 in the complex plane. We can see this in terms of cartesian coordinates. If $z = x + iy$, then $z - i = x + i(y - 1)$ and $z - 3 = x - 3 + iy$, so that z belongs to A if and only if

$$x^2 + (y - 1)^2 = (x - 3)^2 + y^2 \,.$$

Simplifying, this becomes $y = 3x - 4$, the equation of a straight line.

(3) The set $\{\, z : \operatorname{Im} z > 0 \,\}$ is the set of those complex numbers $z = x + iy$ such that $\operatorname{Im} z = y > 0$. This is just the set of all points in the upper half-plane—those points lying *strictly above* the x-axis.

Proposition 1.2   *For any complex number z,*

$$|\operatorname{Re} z| \leq |z|$$

$$|\operatorname{Im} z| \leq |z|$$

$$\tfrac{1}{\sqrt{2}}\left(|x| + |y|\right) \leq |z| \leq |x| + |y|$$

where $x = \operatorname{Re} z$ and $y = \operatorname{Im} z$.

Proof.   The first two inequalities are direct consequences of the inequality $a^2 \leq a^2 + b^2$, valid for any real numbers a and b (take positive square roots).

Furthermore, $|z|^2 = x^2 + y^2 \leq x^2 + y^2 + 2\,|x|\,|y| = (|x| + |y|)^2$. Taking positive square roots gives $|z| \leq |x| + |y|$.

Finally, the inequality $(a - b)^2 \geq 0$, for any real numbers a and b, can be rewritten as $2ab \leq a^2 + b^2$. Using this, we have

$$\left(|x| + |y|\right)^2 = |x|^2 + |y|^2 + 2\,|x|\,|y| \leq 2\,|x|^2 + 2\,|y|^2.$$

Taking the positive square root completes the proof.   □

1.3   Properties of the Modulus

Further properties of the modulus are as follows.

Proposition 1.3   *For any complex numbers z, ζ, we have*

(i) $|z| = 0$ *if and only if $z = 0$;*

(ii) $|z|^2 = z\,\overline{z}$;

(iii) $|z| = |\overline{z}|$;

(iv) $|z\,\zeta| = |z|\,|\zeta|$;

(v)   *if $\zeta \neq 0$, then $\left|\dfrac{z}{\zeta}\right| = \dfrac{|z|}{|\zeta|}$* ;

(vi) $|z + \zeta| \leq |z| + |\zeta|$ (the triangle inequality) ;

(vii) $|z_1 + \cdots + z_m| \leq |z_1| + \cdots + |z_m|$ *for any $z_1, z_2, \ldots, z_m \in \mathbb{C}$.*

Proof.   Parts (i), (ii) and (iii) are straightforward. To prove (iv), we observe that $|z\,\zeta|^2 = z\zeta\,\overline{z\zeta}$, by (ii), $= z\,\zeta\,\overline{z}\,\overline{\zeta} = |z|^2\,|\zeta|^2$, again by (ii).

Taking positive square roots gives (iv). Similarly, if $\zeta \neq 0$, then

$$\left|\frac{z}{\zeta}\right|^2 = \frac{z}{\zeta}\overline{\left(\frac{z}{\zeta}\right)} = \frac{z\bar{z}}{\zeta\bar{\zeta}} = \frac{|z|^2}{|\zeta|^2}$$

and (v) follows.

It is possible to prove part (vi) by substituting in the real and imaginary parts and doing a bit of algebra. However, we can give a slick and relatively painless proof as follows:

$$
\begin{aligned}
|z + \zeta|^2 &= \overline{(z + \zeta)}\,(z + \zeta) \\
&= (\bar{z} + \bar{\zeta})\,(z + \zeta) \\
&= \bar{z}z + \bar{\zeta}z + \bar{z}\zeta + \bar{\zeta}\zeta \\
&= |z|^2 + |\zeta|^2 + z\bar{\zeta} + \overline{z\bar{\zeta}} \\
&= |z|^2 + |\zeta|^2 + 2\,\mathrm{Re}(z\bar{\zeta}) \\
&\leq |z|^2 + |\zeta|^2 + 2\,|z\bar{\zeta}| \\
&= |z|^2 + |\zeta|^2 + 2\,|z|\,|\zeta| \\
&= (|z| + |\zeta|)^2.
\end{aligned}
$$

Taking positive square roots completes the proof.

Part (vii) is the generalized triangle inequality and follows directly from part (vi) by induction. Indeed, for each $m \in \mathbb{N}$, let $P(m)$ be the statement that $|z_1 + \cdots + z_m| \leq |z_1| + \cdots + |z_m|$ for any $z_1, z_2, \ldots, z_m \in \mathbb{C}$. Clearly, $P(1)$ is true.

We suppose that $P(n)$ is true and show that this implies that $P(n+1)$ is true. Indeed, for any $z_1, \ldots, z_{n+1}$ in $\mathbb{C}$, let $\zeta = z_n + z_{n+1}$. Then, we have

$$
\begin{aligned}
|z_1 + \cdots + z_n + z_{n+1}| &= |z_1 + \cdots + z_{n-1} + \zeta| \\
&\leq |z_1| + \cdots + |z_{n-1}| + |\zeta|
\end{aligned}
$$

by the induction hypothesis (namely, that $P(n)$ is true)

$$\leq |z_1| + \cdots + |z_{n-1}| + |z_n| + |z_{n+1}|,$$

by part (vi), and therefore $P(n+1)$ is true, as claimed. Hence, by induction, $P(n)$ is true for all $n \in \mathbb{N}$. $\qquad\square$

Remark 1.1   Replacing ζ by $-\zeta$, the triangle inequality becomes

$$|z - \zeta| \le |z| + |\zeta|\,.$$

Now, $|z| = |z - 0|$ and $|\zeta| = |\zeta - 0|$, and so the above inequality tells us that the distance between the pair of complex numbers z and ζ is no greater than the the sum of the distances of each of z and ζ from the origin. This is just the statement that if we form the triangle with vertices 0, z and ζ, then the length of the side joining z and ζ is never longer than the sum of the other two sides—hence the name "triangle inequality".

For any complex numbers u, v, w, we see that

$$|u - w| = |(u - v) + (v - w)| \le |u - v| + |v - w|\,,$$

so there is nothing special about the origin in the above discussion—it works for any triangle.

Evidently, part (vi) is just a special but important case of part (vii), the generalized triangle inequality.

Remark 1.2   For any $z, \zeta \in C$, we have $|\zeta| = |\zeta - z + z| \le |\zeta - z| + |z|$, giving $|\zeta| - |z| \le |z - \zeta|$. Interchanging z and ζ, we obtain the inequality $|z| - |\zeta| \le |\zeta - z| = |z - \zeta|$. Thus $-|z - \zeta| \le |z| - |\zeta| \le |z - \zeta|$, which can be written as

$$\big|\,|z| - |\zeta|\,\big| \le |z - \zeta|\,.$$

From this, we see that if two complex numbers z and ζ are close (i.e., the distance between them, $|z - \zeta|$, is small) then they have nearly the same modulus. The converse, however, need not be true, for example, i and $-i$ have the same modulus, namely 1, but $|i - (-i)| = |2i| = 2$.

Example 1.2   If the complex numbers u and v are proportional, with positive constant of proportionality, then $u + v = u + ru = (1 + r)u$ for some $r > 0$. Evidently, $|u + v| = (1 + r)|u| = |u| + |v|$. Geometrically, this is clear. The complex number $u + v$ is got by putting the vector v onto the end of the vector u in the complex plane. If $v = ru$, then u and v "line up" and the triangle with vertices 0, u and $u + v$ collapses to a straight line.

Furthermore, if $v_1, \ldots, v_m$ are each of the form $v_j = r_j u$, for some $r_j > 0$, then the vector $u + v_1 + v_2 + \cdots + v_m$ is got by placing parallel vectors end to end and so its length will be the sum of the parts,

$$|u + v_1 + v_2 + \cdots + v_m| = |u| + |v_1| + \cdots + |v_m|\,.$$

We shall see that the converse is also true, as one might expect.

First, we shall show that the equality $|z_1 + z_2| = |z_1| + |z_2|$ holds for non-zero complex numbers, z_1 and z_2, only if they are proportional (with positive constant of proportionality), that is, if and only if $z_2 = rz_1$ for some real number $r > 0$.

Indeed, as we have just discussed, if $z_2 = rz_1$, with $r > 0$, then the claimed equality holds.

Conversely, suppose that $|z_1 + z_2| = |z_1| + |z_2|$. Then

$$\left(|z_1| + |z_2|\right)^2 = |z_1 + z_2|^2 = |z_1|^2 + |z_2|^2 + 2\operatorname{Re}(z_1 \bar{z}_2)$$

and so $\operatorname{Re}(z_1 \bar{z}_2) = |z_1||z_2| = |z_1 \bar{z}_2|$. It follows that $z_1 \bar{z}_2$ has no imaginary part and so $z_1 \bar{z}_2 = \operatorname{Re}(z_1 \bar{z}_2) = |z_1 \bar{z}_2|$. From this we see that $z_2 = r\, z_1$ where $r = |z_1 \bar{z}_2| / |z_1|^2$.

Now, suppose that $z_k \neq 0$, for all $k = 1, \ldots, n$, and that

$$\sum_{k=1}^{n} |z_k| = \left| \sum_{k=1}^{n} z_k \right|. \qquad (*)$$

We wish to show that there are positive real numbers $r_2, \ldots, r_n$ such that $z_j = r_j z_1$ for $j = 2, \ldots, n$. Now, for any partition of the set $\{1, 2, \ldots, n\}$ into two subsets I and J, the equality $(*)$ implies that

$$\sum_{k=1}^{n} |z_k| = \left| \sum_{k=1}^{n} z_k \right| = \left| \sum_{k \in I} z_k + \sum_{k \in J} z_k \right|$$

$$\leq \left| \sum_{k \in I} z_k \right| + \left| \sum_{k \in J} z_k \right|$$

$$\leq \left| \sum_{k \in I} z_k \right| + \sum_{k \in J} |z_k|$$

$$\leq \sum_{k \in I} |z_k| + \sum_{k \in J} |z_k| = \sum_{k=1}^{n} |z_k|$$

and so $\left| \sum_{k \in I} z_k \right| = \sum_{k \in I} |z_k|$. Taking $I = \{1, j\}$ and applying the first part, the result follows.

Let $\zeta_0 = 0$ and $\zeta_j = \zeta_{j-1} + z_j$, for $j = 1, \ldots, n$ and let P be the polygon $[\zeta_0, \zeta_1] \cup \cdots \cup [\zeta_{n-1}, \zeta_n]$, where $[w, z]$ denotes the straight line segment from w to z. Then the equality in question is the statement that the distance between the initial and final points of P is equal to the sum of the lengths of its segments. This can only happen if the polygon stretches out into a straight line (and does not turn back on itself).

Example 1.3   For given $z_1, z_2, z_3 \in \mathbb{C}$, what is the set

$$\{\, w \in \mathbb{C} : w = \alpha z_1 + \beta z_2 + \gamma z_3, \text{ some } \alpha, \beta, \gamma \in [0,1] \text{ with } \alpha + \beta + \gamma = 1 \,\}?$$

In fact, this set is the triangle (including its interior) with vertices z_1, z_2, z_3. To see this, first notice that we can write

$$w = \alpha z_1 + \beta z_2 + \gamma z_3 = (1-\gamma)\big((1-\mu)z_1 + \mu z_2\big) + \gamma z_3$$

where $\mu = \beta/(\alpha+\beta)$ (assuming that the denominator is not zero).] Now, the set $\{\, \zeta : \zeta = (1-\mu)z_1 + \mu z_2, \ 0 \le \mu \le 1 \,\}$ is just the line segment from z_1 to z_2. Let $\zeta_\mu = (1-\mu)z_1 + \mu z_2$ be some point on this line segment. The set $L_\mu = \{\, w : w = (1-\gamma)\zeta_\mu + \gamma z_3, \ 0 \le \gamma \le 1 \,\}$ is the line segment from ζ_μ to z_3. As μ varies between 0 and 1, so ζ_μ varies along the line segment from z_1 to z_2 and the L_μs fill out the triangle.

1.4   The Argument of a Complex Number

We have agreed that a complex number can be usefully pictured as a point in the plane. Now, we can use polar coordinates rather than cartesian coordinates, giving the correspondences (assuming $z \neq 0$)

$$z = x + iy \quad \longleftrightarrow \quad (x,y) \quad \longleftrightarrow \quad (r,\theta),$$

where $r = \sqrt{x^2 + y^2} = |z|$, and where θ is given by the pair of equations

$$\cos\theta = \frac{x}{r} = \frac{x}{|z|} = \frac{\operatorname{Re} z}{|z|}$$

and

$$\sin\theta = \frac{y}{r} = \frac{y}{|z|} = \frac{\operatorname{Im} z}{|z|}.$$

The value of θ is determined only to within additive multiples of 2π, that is, if θ satisfies both $\cos\theta = x/r$ and $\sin\theta = y/r$ then so does $\theta + 2k\pi$, for any $k \in \mathbb{Z}$. Moreover, these are the only possibilities: if ψ also satisfies $\cos\psi = x/r$ and $\sin\psi = y/r$ then $\psi = \theta + 2n\pi$ for some suitable $n \in \mathbb{Z}$.

The angle θ is called the argument of z, denoted $\arg z$. Note that according to the above discussion, $\arg z$ is not well-defined. One could call $\arg z$ an argument of z, i.e., a solution to $\cos\theta = x/r$ and $\sin\theta = y/r$, or one could define $\arg z$ to be the *set* of all such solutions, $\arg z = \{\, \theta, \theta \pm 2\pi, \theta \pm 4\pi, \cdots \}$,

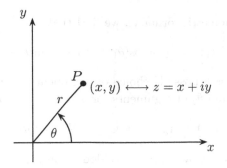

Fig. 1.2   Cartesian and polar coordinates.

where θ is any solution. We prefer the first idea, even though it is something of a nuisance.

By convention, we can pick on a particular choice. There is a unique solution θ satisfying $-\pi < \theta \leq \pi$; this choice of θ is called the principal value of the argument of the complex number z and is denoted by $\text{Arg}\, z$. Thus, for any $z \neq 0$, $\text{Arg}\, z$ *is* well-defined and is uniquely determined by the requirement that $\text{Arg}\, z = \theta \in (-\pi, \pi]$ and $\cos\theta = x/r$ and $\sin\theta = y/r$. For example, $\text{Arg}\, x = 0$ for any real number x with $x > 0$. If x is real and $x < 0$, then $\text{Arg}\, x = \pi$. Also $\text{Arg}\, i = \pi/2$, $\text{Arg}(-1) = \pi$, $\text{Arg}(-i) = -\pi/2$.

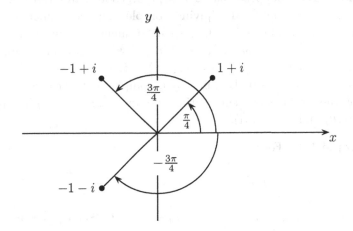

Fig. 1.3 $\text{Arg}(1+i) = \frac{\pi}{4}$, $\text{Arg}(-1+i) = \frac{3\pi}{4}$, $\text{Arg}(-1-i) = -\frac{3\pi}{4}$.

If $z_1 = r_1(\cos\theta_1 + i\sin\theta_1)$ and $z_2 = r_2(\cos\theta_2 + i\sin\theta_2)$ then, using the standard trigonometric formulae, we find that

$$z_1 z_2 = r_1 r_2(\cos(\theta_1 + \theta_2) + i\sin(\theta_1 + \theta_2))$$

and so $\theta_1 + \theta_2$ is a possible choice of argument for the product $z_1 z_2$, for any choices θ_1 and θ_2 of arguments for z_1 and z_2, respectively. In general,

$$\arg(z_1 z_2) = \arg z_1 + \arg z_2 + 2k\pi, \quad \text{for some } k \in \mathbb{Z},$$

where $\arg z_1$, $\arg z_2$ and $\arg(z_1 z_2)$ denote any particular choices of the arguments. Of course, different choices will lead to different values for k.

In particular, by induction, we obtain De Moivre's formula

$$(\cos\theta + i\sin\theta)^n = \cos n\theta + i\sin n\theta.$$

Remark 1.3 It is *not always* true that

$$\text{Arg } z_1 z_2 = \text{Arg } z_1 + \text{Arg } z_2.$$

For example, $\text{Arg}(-1) = \pi$ but $\text{Arg}((-1)(-1)) = \text{Arg } 1 = 0 \neq \pi + \pi$.

The ambiguity of the argument of a complex number reappears when we try to set up the notion of the logarithm of a complex number.

Remark 1.4 Suppose that z has polar coordinates (r, θ), $z \leftrightarrow (r, \theta)$. Then $z^n \leftrightarrow (r^n, n\theta)$. Multiplying complex numbers amounts to multiplying their moduli and adding their arguments. Furthermore, $1/z = \bar{z}/(z\bar{z}) = \bar{z}/|z|^2$, so that $1/z \leftrightarrow (1/r, \varphi)$, where φ is an argument of $\bar{z}$. But $\bar{z} = r\cos\theta - ir\sin\theta = r(\cos(-\theta) + i\sin(-\theta))$ giving $\bar{z} \leftrightarrow (r, -\theta)$. Hence $1/z \leftrightarrow (1/r, -\theta)$. Dividing by a complex number amounts to dividing by the modulus and subtracting the angle; $z/w \leftrightarrow (r/\rho, \theta - \alpha)$, where $z \leftrightarrow (r, \theta)$ and $w \leftrightarrow (\rho, \alpha)$.

Example 1.4 For any $w \neq 1$,

$$1 + w + w^2 + \cdots + w^{n-1} = \frac{1 - w^n}{1 - w}$$

(because $(1 - w)(1 + w + \cdots + w^{n-1}) = 1 - w^n$). Setting $w = \cos\theta + i\sin\theta$ (so that $w^k = \cos k\theta + i\sin k\theta \leftrightarrow (1, k\theta)$), we find that

$$S_n \equiv \cos\theta + \cos 2\theta + \cdots + \cos n\theta = \frac{\sin((n + \frac{1}{2})\theta) - \sin(\frac{\theta}{2})}{2\sin(\frac{\theta}{2})}$$

and that

$$T_n \equiv \sin\theta + \sin 2\theta + \cdots + \sin n\theta = \frac{\cos(\frac{\theta}{2}) - \cos((n + \frac{1}{2})\theta)}{2\sin(\frac{\theta}{2})}$$

provided $\theta \neq 2k\pi$ for any $k \in \mathbb{Z}$.

Indeed, if we let $\zeta = \cos(\frac{\theta}{2}) - i\sin(\frac{\theta}{2}) \leftrightarrow (1, -\frac{\theta}{2})$, then we have

$$\begin{aligned}
S_n + iT_n &= \cos\theta + i\sin\theta + \cos 2\theta + i\sin 2\theta + \cdots + \cos n\theta + i\sin n\theta \\
&= w + w^2 + \cdots + w^n \\
&= \frac{w - w^{n+1}}{1 - w} \\
&= \frac{(w^{n+1} - w)\,\zeta}{(w - 1)\,\zeta} \\
&= \frac{w^{n+1}\,\zeta - w\,\zeta}{w\,\zeta - \zeta} \\
&= \frac{\cos((n + \frac{1}{2})\theta) + i\sin((n + \frac{1}{2})\theta) - \cos(\frac{\theta}{2}) - i\sin(\frac{\theta}{2})}{2\,i\sin(\frac{\theta}{2})}.
\end{aligned}$$

Equating real and imaginary parts gives the required formulae.

Returning to the general discussion, we have declared a complex number to be one of the form $z = x + iy$, where $i^2 = -1$, and have so far accepted this uncritically. The question is "is this legitimate?" What is this magical number which we denote by i? It is certainly not a real number, so what is it? Now watch closely:

$$\frac{(-1)}{1} = -1 = \frac{1}{(-1)}$$

$$\implies \quad \sqrt{\frac{-1}{1}} = \sqrt{\frac{1}{-1}}$$

$$\implies \quad \frac{\sqrt{-1}}{\sqrt{1}} = \frac{\sqrt{1}}{\sqrt{-1}}$$

$$\implies \quad \sqrt{-1}\,\sqrt{-1} = \sqrt{1}\,\sqrt{1}$$

$$\implies \quad -1 = 1,$$

which is some cause for concern.

A similar but somewhat less picturesque "observation" is that

$$\sqrt{10} = \sqrt{(-5)(-2)} = \sqrt{-5}\,\sqrt{-2} = i\sqrt{5}\ i\sqrt{2} = i^2\sqrt{10} = -\sqrt{10}.$$

Maybe we should take a little more care in setting up the notion of a complex number. Happily, this can be done, as we will now see (but the paradoxes above must wait until we discuss complex powers).

1.5   Formal Construction of Complex Numbers

We define $\mathbb{C}$ to be the set of ordered pairs (x, y), with $x, y \in \mathbb{R}$, together with the binary operations of "addition" (denoted $+$) and "multiplication" (denoted $\cdot$) given, respectively, by

$$(a, b) + (c, d) = (a + c, b + d)$$

and

$$(a, b) \cdot (c, d) = (ac - bd, ad + bc).$$

(Secretly, we think of (a, b) and (c, d) as being $a + ib$ and $c + id$. Then the addition and multiplication laws above are the obvious ones.)

Proposition 1.4   *The set $\mathbb{C}$ equipped with these operations is a field in which $(0, 0)$ is the identity for addition and $(1, 0)$ is the identity for multiplication.*

Proof.   It is clear from the definitions above that $(a, b) + (0, 0) = (a, b) = (0, 0) + (a, b)$ and that $(a, b) \cdot (1, 0) = (a, b) = (1, 0) \cdot (a, b)$. Also, $(-a, -b)$ is an additive inverse for (a, b). Furthermore, provided $(a, b) \neq (0, 0)$, we see that (c, d) is a multiplicative inverse for (a, b) where $c = a/\sqrt{a^2 + b^2}$ and $d = -b/\sqrt{a^2 + b^2}$. (These are what we would expect if (a, b) is to somehow be a rigorous realization of the expression $a + ib$).

Straightforward computations, using the definitions of $+$ and $\cdot$, show that, for any (a, b), (c, d), $(e, f) \in \mathbb{C}$,

$$(a, b) + (c, d) = (c, d) + (a, b), \quad (\, + \text{ is commutative}),$$
$$(a, b) + \big((c, d) + (e, f)\big) = \big((a, b) + (c, d)\big) + (e, f), \quad (\, + \text{ is associative}),$$
$$(a, b) \cdot (c, d) = (c, d) \cdot (a, b), \quad (\, \cdot \text{ is commutative}),$$
$$(a, b) \cdot \big((c, d) \cdot (e, f)\big) = \big((a, b) \cdot (c, d)\big) \cdot (e, f), \quad (\, \cdot \text{ is associative}),$$
$$(a, b) \cdot \big((c, d) + (e, f)\big) = (a, b) \cdot (c, d) + (a, b) \cdot (e, f),$$
$$(\, \cdot \text{ is distributive over } +\,)$$

Thus, $\mathbb{C}$ is a field, as claimed. $\qquad\qquad\qquad\qquad\qquad\qquad\qquad\qquad\square$

Proposition 1.5   *The set* $\mathbb{F} = \{(a, 0) : a \in \mathbb{R}\}$ *is a subfield of* $\mathbb{C}$ *and the map* $\phi : \mathbb{R} \to \mathbb{C}$ *given by* $\phi : a \mapsto (a, 0)$ *is a field isomorphism of* $\mathbb{R}$ *onto* $\mathbb{F}$.

Proof.   From the definitions, we see that $(a, 0) + (b, 0) = (a+b, 0)$ and that $(a, 0) \cdot (b, 0) = (ab, 0)$, for any $(a, 0), (b, 0) \in \mathbb{F}$. Furthermore, the additive inverse of $(a, 0)$ is $(-a, 0) \in \mathbb{F}$ and, if $a \neq 0$, the multiplicative inverse is $(1/a, 0) \in \mathbb{F}$. It follows that $\mathbb{F}$ is a subfield of $\mathbb{C}$.

Next, we note that $\phi(a+b) = (a+b, 0) = (a, 0) + (b, 0) = \phi(a) + \phi(b)$ and $\phi(ab) = (ab, 0) = (a, 0) \cdot (b, 0) = \phi(a) \cdot \phi(b)$, $\phi(0) = (0, 0)$ and $\phi(1) = (1, 0)$ and so ϕ is a homomorphism with respect to both operations $+$ and $\cdot$.

Finally, we observe that $(a, 0) = \phi(a)$ and so ϕ maps $\mathbb{R}$ onto $\mathbb{F}$, and if $\phi(a) = \phi(b)$, then $(a, 0) = (b, 0)$ and therefore $a = b$. Hence ϕ maps $\mathbb{R}$ one-one onto $\mathbb{F}$ and is a field isomorphism.   □

This means that $\mathbb{F}$ and $\mathbb{R}$ are "the same", that is, $\mathbb{R}$ can be embedded in $\mathbb{C}$ as $\mathbb{F}$. This is just the formal proof that the "real line" is still the "real line" when we consider it as the x-axis of the complex plane. This is not an entirely vacuous statement because we are also considering the additive and multiplicative structures involved. (The plane is more naturally considered as a linear space, so that addition is natural but multiplication is a little special. In fact, it can be shown that $\mathbb{R}^n$ (with $n > 1$) can be given a multiplication making it into a field only for $n = 2$, in which case the multiplication is as above.)

Now, any $(a, b) \in \mathbb{C}$ can be written as

$$
\begin{aligned}
(a, b) &= (a, 0) + (0, b) = (a, 0) + (0, 1) \cdot (b, 0) \\
&= \phi(a) + (0, 1) \cdot \phi(b) = \phi(a) + i\,\phi(b) \\
&= a + ib
\end{aligned}
$$

where we have dropped the isomorphism notation ϕ by writing $\phi(x)$ as just x, for any $x \in \mathbb{R}$. Also, we have set $i = (0, 1)$, and we have dropped the $\cdot$, denoting multiplication merely by juxtaposition, as usual. Thus, with this new streamlined notation, any complex number has the form $a + ib$, with $a, b \in \mathbb{R}$, and where i satisfies $i^2 = (0, 1) \cdot (0, 1) = (-1, 0) = \phi(-1) = -1$, i.e., $i^2 = -1$. We have therefore given substance to the hopeful but vague idea of "a number of the form $x + iy$, with $x, y \in \mathbb{R}$, and with $i^2 = -1$", and have recovered our original formulae for addition and multiplication. The complex numbers are well-defined—they form a field and contain the set of real numbers $\mathbb{R}$ as a subfield.

What's going on ?   We might worry about simply asserting that $i^2 = -1$ without having said precisely what i was in the first place. However, it turns out to be all right. We can just go ahead and write any complex number as $a + ib$, where $i^2 = -1$ and not worry. It can all be justified. (But we still need to sort out square roots.)

Remark 1.5   The set of real numbers has a notion of order, defined in terms of positivity. For any real number x, precisely one of the following three statements is true; $x = 0$, $x > 0$, $-x > 0$. A property of positivity is that if $x > 0$ and $y > 0$, then $xy > 0$. It follows that $1 > 0$. (To see this, first we note that, clearly, $1 \neq 0$ (otherwise, $x = 1\,x$ would be 0 for all $x \in \mathbb{R}$). Hence either $1 > 0$ or $-1 > 0$, but not both. If $-1 > 0$ were true, then we would have $(-1)(-1) > 0$. But $(-1)(-1) = 1^2 = 1$ so that also $1 > 0$. Both $-1 > 0$ and $1 > 0$ is not allowed, so we conclude that $-1 > 0$ is false and therefore $1 > 0$.)

Is there such a notion for complex numbers which extends that for the real numbers? If this were possible then, for example, we would have either $i > 0$ or $-i > 0$ (since $i \neq 0$). If $i > 0$ were true, we would have $-1 = i^2 > 0$, which is false. Hence $-i > 0$ must be true. But then, again, this would imply $-1 = (-i)^2 > 0$, which is false. We must concede that there is no generalization of positivity extending from the real numbers to the set of complex numbers.

For real numbers x and y, the inequality $x > y$ is just a way of writing $x - y > 0$. This latter does not make sense, in general, for complex numbers, so it follows that inequalities, such as $z > \zeta$, do not make sense for complex numbers.

What's going on ?   Try as we might, we cannot make (useful) sense of inequalities between complex numbers.

1.6   The Riemann Sphere and the Extended Complex Plane

Let S^2 denote the sphere $\{(x, y, z) \in \mathbb{R}^3 : x^2 + y^2 + z^2 = 1\}$ in $\mathbb{R}^3$ and let N denote the point $(0, 0, 1)$, the "north pole" of S^2. Think of $\mathbb{C}$ as the plane $\{(x, y, z) : z = 0\}$, containing the equator of S^2. Then given any point P in this plane, the straight line through P and N cuts the sphere S^2 in a unique point, P', say. As P varies over the plane, the corresponding point P' varies over $S^2 \setminus \{N\}$. This sets up a one-one correspondence between $\mathbb{C}$ and $S^2 \setminus \{N\}$.

We note that points far from the origin in $\mathbb{C}$ are mapped into points near the north pole (and points close to the origin are mapped into points close to the south pole of S^2, i.e., the point $(0,0,-1)$). Notice too that if (P_n) is a sequence of points in $\mathbb{C}$ which converges to some point P in $\mathbb{C}$, then the images P'_n of P_n converge in S^2 to the image P' of P. We also see that if (z_n) is a sequence in $\mathbb{C}$ such that $|z_n| \to \infty$, then the sequence (P'_n) of their images converges to N in S^2 (and vice versa). The point N is called the "point at infinity".

The extended complex plane, $\mathbb{C}_\infty$, is defined to be $\mathbb{C}$ together with one additional element, that is, $\mathbb{C}_\infty \equiv \mathbb{C} \cup \{\infty\}$, where $\{\infty\}$ is a singleton set with $\infty \notin \mathbb{C}$. It does not matter what ∞ actually is, as long as it is not already a member of the set $\mathbb{C}$. For example, we could take ∞ to be $\varnothing$, which is certainly not a complex number. (Note that a and $\{a\}$ are different mathematical objects, so, in particular, $\varnothing$ is not the same as $\{\varnothing\}$. Indeed, the objects $\{\varnothing\}$, $\{\varnothing, \{\varnothing\}\}$ and $\{\varnothing, \{\varnothing, \{\varnothing\}\}\}$ are different, as different as the numbers $1, 2, 3$.)

What's going on? The issue is how to augment a given set to give it just one new element. That is, given a set A, how does one construct a new set B such that $B \setminus A$ is a singleton set? In the case above, $A = \mathbb{C}$ and $B = \mathbb{C}_\infty$ is the set we seek. If it does not matter what the new element is, as in the case here, then the explicit construction above is just one of many possibilities.

There is then a one-one correspondence between $\mathbb{C}_\infty$ and S^2 given by $\infty \longleftrightarrow N$ together with the correspondence between $\mathbb{C}$ and $S^2 \setminus \{N\}$, as introduced above. The extended complex plane, $\mathbb{C}_\infty$, viewed in this way is referred to as the Riemann sphere.

This gives a sensible realization of "infinity". For example, the mapping $z \mapsto 1/z$ is not *a priori* defined at $z = 0 \in \mathbb{C}$. However, if we consider the extended complex plane, or the Riemann sphere, then in addition to the mapping $z \mapsto 1/z$ for $z \in \mathbb{C} \setminus \{0\}$, we can define $0 \mapsto N$ and $N \mapsto 0$, (or, in more suggestive notation, $0 \mapsto \infty$ ($1/0 \equiv \infty$) and $\infty \mapsto 0$ ($1/\infty \equiv 0$)). This defines the map $z \mapsto 1/z$ as a mapping from $\mathbb{C}_\infty \to \mathbb{C}_\infty$. This construction is reasonable in that if $z_n \to z$, then $1/z_n \to 1/z$ even if z or any z_n is equal to 0 or to ∞.

The point here is to notice that by studying $\mathbb{C}_\infty$, rather than just $\mathbb{C}$, we can sometimes handle singularities just as ordinary points—after all, one point on a sphere is much the same as any other.

In real analysis, one considers the limits $x \to \infty$ and $x \to -\infty$. Whilst it must be stressed at the outset that this is just shorthand symbolism, nevertheless, it does invoke a kind of image of two infinities—one positive and the other negative. In view of the picture of complex numbers as points in the plane, one might wonder if it might be worth considering some kind of collection of "complex infinities", each being somewhere off in some given direction (perhaps corresponding to some "end of the rainbow" at the "end" of the ray $r(\cos\theta + i\sin\theta)$ as r becomes very large). The view of $\mathbb{C}$ as being wrapped around a sphere, as developed above, suggests that we can bundle all these "infinities" into just a single "point at infinity", namely, the north pole.

It should be stressed that whilst $\mathbb{C}$ is a field (so one can do arithmetic), this is no longer true of $\mathbb{C}_\infty$. There is no attempt to assign any meaning whatsoever to expressions such as $\infty + \infty$ or $0 \times \infty$. The operations of addition and multiplication are simply not directly applicable when ∞ is involved.

Chapter 2

Sequences and Series

2.1 Complex Sequences

A sequence of complex numbers is a collection of elements of $\mathbb{C}$ labelled by integers; for example, $a_1, a_2, a_3, \ldots$. To be more precise, a sequence of complex numbers is a map a from the natural numbers $\mathbb{N}$ into $\mathbb{C}$. If we write a_n for the value $a(n)$, then we recover the intuitive notion above. The important thing about a sequence is that there is a notion of "further along", e.g., the term a_{710} is "further along" the sequence than, say, a_{106}. Of course, this property is inherited from the ordering within $\mathbb{N}$; a_n is further along the sequence than a_m if and only if $n > m$.

We denote the sequence $a_1, a_2, \ldots$ by (a_n) or $(a_n)_{n \in \mathbb{N}}$. It is often very convenient to allow a sequence to begin with a_0 rather than with a_1. (We can still express this in terms of a map from $\mathbb{N}$ into $\mathbb{C}$ by considering the sequence (b_n) where $b : \mathbb{N} \to \mathbb{C}$ is the map $b : n \mapsto a_{n-1}$, $n \in \mathbb{N}$.) The notation $(a_n)_{n=0}^{\infty}$ or $(a_n)_{n \geq 0}$ might be appropriate here.

2.2 Subsequences

A subsequence of a sequence (a_n) is any sequence got by removing terms from the original sequence (a_n). For example, $a_1, a_3, a_5, a_7, \ldots$ (where "$\ldots$" means "and all further terms with an odd index") is a subsequence of the sequence $a_1, a_2, a_3, \ldots$. More formally, we define subsequences via mappings on $\mathbb{N}$ as follows. Suppose that $\phi : \mathbb{N} \to \mathbb{C}$ is a given sequence in $\mathbb{C}$. Let $\psi : \mathbb{N} \to \mathbb{N}$ be any given map such that $\psi(n) > \psi(m)$ whenever $n > m$ (i.e., ψ preserves the order in $\mathbb{N}$). Then $\phi \circ \psi : \mathbb{N} \to \mathbb{C}$ is a sequence of complex numbers; $\phi \circ \psi$ maps n into $\phi(\psi(n)) \in \mathbb{C}$, for $n \in \mathbb{N}$. Any sequence of this form is said to be a subsequence of ϕ.

In our example above, suppose that $\phi(n) = a_n$. Let $\psi : \mathbb{N} \to \mathbb{N}$ be the map $\psi : k \mapsto 2k - 1$, for $k \in \mathbb{N}$ (note that ψ preserves the order in $\mathbb{N}$). Then the sequence $\phi \circ \psi$ is $n \mapsto \phi(\psi(n)) = \phi(2n - 1) = a_{2n-1} = a_{\psi(n)}$, i.e., the subsequence $a_{\psi(1)}, a_{\psi(2)}, a_{\psi(3)}, \dots = a_1, a_3, a_5, \dots$. Often the image $\psi(k)$ of $k \in \mathbb{N}$ under ψ is denoted by n_k, in which case the subsequence $\phi \circ \psi$ is denoted by $(a_{n_k})_{k \in \mathbb{N}}$ (or $(a_{n_k})_{k \geq 0}$ if we start with $k = 0$ rather than with $k = 1$).

2.3 Convergence of Sequences

We have a useful notion of distance between two complex numbers, so we can use this to define convergence in $\mathbb{C}$.

Definition 2.1 The sequence (a_n) of complex numbers converges to ζ in $\mathbb{C}$ if for any given $\varepsilon > 0$ there is $N \in \mathbb{N}$ such that $|a_n - \zeta| < \varepsilon$ whenever $n > N$. ζ is the limit of the sequence.

We signify that (a_n) converges to ζ by writing $a_n \to \zeta$, as $n \to \infty$, or $\lim_{n \to \infty} a_n = \zeta$.

Remarks 2.1

(1) One expects that the smaller the given ε, the larger N will need to be.
(2) This definition looks exactly the same, typographically, as that of the convergence of a sequence of real numbers. In fact, if ζ is real and each a_n is real, then this reduces to the definition for real numbers. In other words, this is a generalization of the notion for real sequences to complex ones.
(3) The value $|a_n - \zeta|$ is the distance between a_n and ζ, and so (a_n) converges to ζ if and only if the distances between ζ and the various terms a_n become smaller than any preassigned positive value provided we go far enough "along the sequence". This last part is usefully paraphrased by saying that for any given $\varepsilon > 0$, the distance between ζ and a_n is "eventually" smaller than ε.

Our first result confirms that any subsequence of a convergent sequence also converges, and to the same limit as the original sequence.

Proposition 2.1 *Suppose $a_n \to \zeta$ as $n \to \infty$ and (a_{n_k}) is a subsequence of (a_n). Then $a_{n_k} \to \zeta$ as $k \to \infty$.*

Proof. Let $\varepsilon > 0$ be given. Since (a_n) converges to ζ, there is some $N \in \mathbb{N}$ such that $n > N$ implies that $|a_n - \zeta| < \varepsilon$. Now, (n_k) is a strictly increasing sequence of integers and so there is K such that $n_K > N$. But then $n_k > n_K > N$ whenever $k > K$. It follows that $|a_{n_k} - \zeta| < \varepsilon$ whenever $k > K$, i.e., by definition, (a_{n_k}) converges to ζ. $\qquad\square$

As a consequence, we can say that a sequence with two convergent subsequences, but with different limits, cannot converge (for if it did, every convergent subsequence would have the same limit, namely, the limit of the original sequence). For example, the sequence $(a_n) = ((-1)^n)$ does not converge, since $a_1, a_3, a_5, \ldots$ and $a_2, a_4, a_6, \ldots$ are convergent subsequences with limits -1 and 1, respectively.

What does it mean to say that a given sequence does not converge? The sequence (b_n) does not converge if it is never eventually close to any point in $\mathbb{C}$; that is, for any given $\zeta \in \mathbb{C}$, (b_n) is not eventually close to ζ. This means that there some $\varepsilon_0 > 0$ such that it is false that (b_n) is eventually within ε_0 of ζ. This, in turn, means that no matter what we choose for N, there is always some $n > N$ such that $|b_n - \zeta| > \varepsilon_0$. In particular, there is n_1 such that $|b_{n_1} - \zeta| > \varepsilon_0$. Similarly, there is $n_2 > n_1$ such that $|b_{n_2} - \zeta| > \varepsilon_0$, and so on, giving an increasing sequence $n_1 < n_2 < n_3 < \cdots$ such that $|b_{n_k} - \zeta| > \varepsilon_0$ for $k = 1, 2, \ldots$.

Hence, for any given $\zeta \in \mathbb{C}$, there is some $\varepsilon_0 > 0$ and some subsequence (b_{n_k}) with the property that $|b_{n_k} - \zeta| > \varepsilon_0$, $k \in \mathbb{N}$. The subsequence and ε_0 will, in general, depend on the chosen point ζ.

The next theorem tells us that the convergence of a sequence of complex numbers is equivalent to that of its real and imaginary parts, with the real and imaginary parts of the limit being the limits, respectively, of the real and imaginary parts of the sequence.

Theorem 2.1 *The sequence of complex numbers (a_n) converges to ζ if and only if both $\operatorname{Re} a_n \to \operatorname{Re} \zeta$ and $\operatorname{Im} a_n \to \operatorname{Im} \zeta$, as $n \to \infty$.*

Proof. Suppose $a_n = \alpha_n + i\beta_n$ and $\zeta = \xi + i\eta$ and that $a_n \to \zeta$ as $n \to \infty$. Let $\varepsilon > 0$ be given. Then there is some $N \in \mathbb{N}$ such that $|a_n - \zeta| < \varepsilon$, whenever $n > N$. The inequalities

$$|\operatorname{Re} a_n - \operatorname{Re} \zeta| = |\operatorname{Re}(a_n - \zeta)| \le |a_n - \zeta|$$

and

$$|\operatorname{Im} a_n - \operatorname{Im} \zeta| = |\operatorname{Im}(a_n - \zeta)| \le |a_n - \zeta|$$

show that both $|\operatorname{Re} a_n - \operatorname{Re} \zeta| < \varepsilon$ and $|\operatorname{Im} a_n - \operatorname{Im} \zeta| < \varepsilon$ whenever $n > N$, that is, $\operatorname{Re} a_n \to \operatorname{Re} \zeta$ and $\operatorname{Im} a_n \to \operatorname{Im} \zeta$ as $n \to \infty$.

For the converse, suppose that both $\operatorname{Re} a_n \to \operatorname{Re} \zeta$ and $\operatorname{Im} a_n \to \operatorname{Im} \zeta$ as $n \to \infty$. Let $\varepsilon > 0$ be given.

Then there is $N_1 \in \mathbb{N}$ such that $|\operatorname{Re} a_n - \operatorname{Re} \zeta| < \varepsilon/2$ whenever $n > N_1$ and there is $N_2 \in \mathbb{N}$ such that $|\operatorname{Im} a_n - \operatorname{Im} \zeta| < \varepsilon/2$ whenever $n > N_2$. Let $N = N_1 + N_2$ (or $\max\{N_1, N_2\}$, it will work just as well). Then $n > N$ implies that

$$|a_n - \zeta| \leq |\operatorname{Re} a_n - \operatorname{Re} \zeta| + |\operatorname{Im} a_n - \operatorname{Im} \zeta| < \frac{\varepsilon}{2} + \frac{\varepsilon}{2} = \varepsilon$$

and the result follows. $\square$

Remark 2.2 A somewhat streamlined proof of this last result can be given as follows. We begin by noting that the convergence of (a_n) to ζ is equivalent to that of the real sequence $(|a_n - \zeta|)$ to 0. This is equivalent to that of $(|a_n - \zeta|^2)$ to 0, i.e., it is equivalent to the convergence of the real sequence $((\alpha_n - \xi)^2 + (\beta_n - \eta)^2)$ to 0 in $\mathbb{R}$. But this, in turn, is equivalent to the convergence of both (α_n) to ξ (in $\mathbb{R}$) and (β_n) to η (in $\mathbb{R}$), and the proof is complete.

Theorem 2.2 *Suppose that $z_n \to z$ and $\zeta_n \to \zeta$ as $n \to \infty$. Then*

(i) $a z_n + b \zeta_n \to a z + b \zeta$, *for any $a, b \in \mathbb{C}$;*
(ii) $z_n \zeta_n \to z \zeta$ *as $n \to \infty$;*
(iii) *if $\zeta_n \neq 0$ for all n and $\zeta \neq 0$, then $1/\zeta_n \to 1/\zeta$ as $n \to \infty$;*
(iv) *if $\zeta_n \neq 0$ for all n and $\zeta \neq 0$, then $z_n/\zeta_n \to z/\zeta$ as $n \to \infty$.*

Proof. We can prove these statements directly or we could appeal to the corresponding familiar statements for real sequences. For example, we can prove (iii) directly as follows. Let $\varepsilon > 0$ be given. We wish to show that eventually $|1/\zeta_n - 1/\zeta| < \varepsilon$. Now, $|1/\zeta_n - 1/\zeta| = |\zeta - \zeta_n| / |\zeta_n \zeta|$ and so we must say something about both terms on the right hand side. We know that $|\zeta - \zeta_n|$ is eventually small because we are told that $\zeta_n \to \zeta$.

We are also told that $\zeta \neq 0$ and so $|\zeta| \neq 0$ as well. Taking $\frac{1}{2}|\zeta|$ as our "$\varepsilon > 0$", we deduce that there is some $N_1 \in \mathbb{N}$ such that $|\zeta - \zeta_n| < \frac{1}{2}|\zeta|$, whenever $n > N_1$. Hence

$$|\zeta| = |\zeta - \zeta_n + \zeta_n| \leq |\zeta - \zeta_n| + |\zeta_n| < \tfrac{1}{2}|\zeta| + |\zeta_n|$$

whenever $n > N_1$, and therefore (rearranging) $\frac{1}{2}|\zeta| < |\zeta_n|$ whenever $n > N_1$. Let $\kappa = \min\{\frac{1}{2}|\zeta|, |\zeta_1|, \ldots, |\zeta_{N_1}|\}$. Then $\kappa > 0$, since each of the $N_1 + 1$ terms is positive, and $|\zeta_n| \geq \kappa$ for all $n \in \mathbb{N}$. Hence $1/(|\zeta_n| |\zeta|) \leq 1/(\kappa |\zeta|)$ for all n.

We are now in a position to piece things together.

Let $\varepsilon > 0$ be given. Then there is $N \in \mathbb{N}$ such that $n > N$ implies that

$$|\zeta_n - \zeta| < \varepsilon \kappa |\zeta| .$$

Note that the right hand side is strictly positive. Hence

$$|1/\zeta - 1/\zeta_n| = \frac{|\zeta_n - \zeta|}{|\zeta_n||\zeta|} < \frac{\varepsilon \kappa |\zeta|}{\kappa |\zeta|} = \varepsilon$$

whenever $n > N$, as required.

Next, we shall prove (ii) via the familiar real versions. To this end, let $z_n = x_n + iy_n$, $z = x + iy$, $\zeta_n = \xi_n + i\eta_n$ and $\zeta = \xi + i\eta$ where each $x_n, y_n, \xi_n, \eta_n, x, y, \xi, \eta$ belongs to $\mathbb{R}$. Then

$$z_n\zeta_n = (x_n + iy_n)(\xi_n + i\eta_n) = x_n\xi_n - y_n\eta_n + i(x_n\eta_n + y_n\xi_n).$$

We are told that $z_n \to z$ and $\zeta_n \to \zeta$ and so $x_n \to x$, $y_n \to y$, $\xi_n \to \xi$ and $\eta_n \to \eta$. By "real analysis", we conclude that $x_n\xi_n - y_n\eta_n \to x\xi - y\eta = \operatorname{Re} z\zeta$ and $x_n\eta_n + y_n\xi_n \to x\eta + y\xi = \operatorname{Im} z\zeta$. In other words, $\operatorname{Re} z_n\zeta_n \to \operatorname{Re} z\zeta$ and $\operatorname{Im} z_n\zeta_n \to \operatorname{Im} z\zeta$, and therefore $z_n\zeta_n \to z\zeta$.

Part (i) can be verified in a similar way (or directly), as can (iv). We also note that (iv) follows from (ii) and (iii). $\square$

2.4 Cauchy Sequences

It makes sense to talk about Cauchy sequences in $\mathbb{C}$, they are defined just as for the real case, as follows.

Definition 2.2 The sequence (z_n) is said to be a Cauchy sequence in $\mathbb{C}$ if for any given $\varepsilon > 0$ there is $N \in \mathbb{N}$ such that $|z_n - z_m| < \varepsilon$ whenever both $m, n > N$.

Proposition 2.2 *The sequence (z_n) is a Cauchy sequence in $\mathbb{C}$ if and only if both $(\operatorname{Re} z_n)$ and $(\operatorname{Im} z_n)$ are Cauchy sequences in $\mathbb{R}$.*

Proof. The proof follows from the observations that

$$\operatorname{Re} z_n - \operatorname{Re} z_m = \operatorname{Re}(z_n - z_m) \quad \text{and} \quad \operatorname{Im} z_n - \operatorname{Im} z_m = \operatorname{Im}(z_n - z_m),$$

so that

$$\left.\begin{array}{l} |\operatorname{Re} z_n - \operatorname{Re} z_m| \\[2ex] |\operatorname{Im} z_n - \operatorname{Im} z_m| \end{array}\right\} \le |z_n - z_m| \le |\operatorname{Re} z_n - \operatorname{Re} z_m| + |\operatorname{Im} z_n - \operatorname{Im} z_m|.$$

The left hand inequalities in each pair imply that both $(\operatorname{Re} z_n)$ and $(\operatorname{Im} z_n)$ are Cauchy sequences if (z_n) is, whereas the right hand inequality implies that (z_n) is a Cauchy sequence whenever both $(\operatorname{Re} z_n)$ and $(\operatorname{Im} z_n)$ are.

(For given $\varepsilon > 0$ there is N_1 such that $|\operatorname{Re} z_n - \operatorname{Re} z_m| < \varepsilon/2$ whenever $m, n > N_1$. Similarly, there is N_2 such that $|\operatorname{Im} z_n - \operatorname{Im} z_m| < \varepsilon/2$ whenever $m, n > N_2$. Putting $N = \max\{N_1, N_2\}$, we find that $|z_n - z_m| \le |\operatorname{Re} z_n - \operatorname{Re} z_m| + |\operatorname{Im} z_n - \operatorname{Im} z_m| < \varepsilon$ whenever $m, n > N$.) $\qquad\square$

The set of complex numbers $\mathbb{C}$, in common with the set of real numbers $\mathbb{R}$, possesses the property that Cauchy sequences necessarily converge (the completeness property). In fact, this important property is inherited from the real number system, as we show next. (It does not hold, for example, for $\mathbb{Q}$, the set of (real) rationals.)

Theorem 2.3 *A sequence of complex numbers (z_n) converges in $\mathbb{C}$ if and only if it is a Cauchy sequence.*

Proof. Suppose that $z_n \to z$ as $n \to \infty$. We shall show that (z_n) is a Cauchy sequence. Let $\varepsilon > 0$ be given. Then there is $N \in \mathbb{N}$ such that $|z_n - z| < \varepsilon/2$ whenever $n > N$. Hence, for any $m, n > N$, we have that

$$\begin{aligned} |z_n - z_m| &= |(z_n - z) + (z - z_m)| \\ &\le |z_n - z| + |z - z_m| \\ &< \varepsilon \end{aligned}$$

and so we see that (z_n) is a Cauchy sequence, as claimed.

Conversely, suppose that (z_n) is a Cauchy sequence in $\mathbb{C}$. Then both $(\operatorname{Re} z_n)$ and $(\operatorname{Im} z_n)$ are Cauchy sequences of real numbers. It is a basic property of the real number system $\mathbb{R}$ that any Cauchy sequence of real numbers has a limit ($\mathbb{R}$ is complete). Hence there is some α in $\mathbb{R}$ and some β in $\mathbb{R}$ such that $\operatorname{Re} z_n \to \alpha$ and $\operatorname{Im} z_n \to \beta$ as $n \to \infty$. But then $z_n \to \alpha + i\beta$ as $n \to \infty$. $\qquad\square$

2.5 Complex Series

We can set up a theory of complex series just as is done for their real counterpart. A series is simply a limit of a sequence of partial sums. Of course, it is necessary to worry about the existence of any such limit, that is, we must pay attention to the convergence, or otherwise, of the sequence of partial sums.

Definition 2.3 Suppose that $w_0, w_1, w_2, \ldots$ is a sequence of complex numbers. The series $\sum_{k=0}^{\infty} w_k$ is said to converge provided that the sequence of partial sums $S_n = \sum_{k=0}^{n} w_k$ converges in $\mathbb{C}$ as $n \to \infty$; otherwise, it is said to diverge. If $S_n \to S$ as $n \to \infty$, then we write $\sum_{k=0}^{\infty} w_k$ for S, the limit of the series.

By considering the real and imaginary parts, we see that the series $\sum_{k=0}^{\infty} w_k$ converges if and only if both series $\sum_{k=0}^{\infty} \operatorname{Re} w_k$ and $\sum_{k=0}^{\infty} \operatorname{Im} w_k$ converge. If this is the case, then

$$\sum_{k=0}^{\infty} w_k = \sum_{k=0}^{\infty} \operatorname{Re} w_k + i \sum_{k=0}^{\infty} \operatorname{Im} w_k.$$

Indeed, for any $n \in \mathbb{N}$, $S_n = \operatorname{Re} S_n + i \operatorname{Im} S_n$ and S_n converges if and only both $\operatorname{Re} S_n$ and $\operatorname{Im} S_n$ converge. Writing $w_k = u_k + i v_k$, we see that $\operatorname{Re} S_n = \operatorname{Re} \sum_{k=0}^{n} w_k = \sum_{k=0}^{n} u_k$ and, similarly, $\operatorname{Im} S_n = \sum_{k=0}^{n} v_k$. Therefore S_n converges if and only if both $\sum_{k=0}^{\infty} u_k$ and $\sum_{k=0}^{\infty} v_k$ converge; in which case

$$\sum_{k=0}^{\infty} w_k = \sum_{k=0}^{\infty} u_k + i \sum_{k=0}^{\infty} v_k,$$

as claimed.

Remark 2.3 If $\sum_{k=0}^{\infty} w_k$ converges, then $S_n = \sum_{k=0}^{n} w_k$ converges and so is a Cauchy sequence; that is, for any given $\varepsilon > 0$, there is $N \in \mathbb{N}$ such that $|S_n - S_m| < \varepsilon$ whenever $m, n > N$. In particular, if $m = n - 1$ and $n > N + 1$ (so that $m > N$), then $|S_n - S_{n-1}| < \varepsilon$. But $S_n - S_{n-1} = w_n$, and so $|w_n| < \varepsilon$ whenever $n > N + 1$; that is $w_n \to 0$ as $n \to \infty$. This observation tells us that if it is *false* that $w_n \to 0$, as $n \to \infty$, then $\sum_{k=0}^{\infty} w_k$ does *not* converge.

Notice that $w_n \to 0$ as $n \to \infty$ is a necessary but *not* a sufficient condition for the convergence of $\sum_{k=0}^{\infty} w_k$. Indeed, taking $w_k = 1/(k+1)$, $k = 0, 1, 2, \ldots$ provides a counterexample (familiar from real analysis).

Proposition 2.3   *Suppose that both the series $\sum_{k=0}^{\infty} w_k$ and $\sum_{k=0}^{\infty} \zeta_k$ are convergent. Then, for any a, $b \in \mathbb{C}$, the series $\sum_{k=0}^{\infty}(aw_k + b\zeta_k)$ converges and*

$$\sum_{k=0}^{\infty}(aw_k + b\zeta_k) = a\sum_{k=0}^{\infty} w_k + b\sum_{k=0}^{\infty} \zeta_k.$$

In other words, convergent series can be added term by term, and multiplication by any complex number can be performed termwise.

Proof.   This follows from $\sum_{k=0}^{n} aw_k + b\zeta_k = a\sum_{k=0}^{n} w_k + b\sum_{k=0}^{n} \zeta_k.$ □

2.6   Absolute Convergence

The following definition of absolute convergence is just as for real series.

Definition 2.4   The series $\sum_{k=0}^{\infty} w_k$ is said to converge absolutely if the series $\sum_{k=0}^{\infty} |w_k|$ (with real non-negative terms) converges.

The series $\sum_{k=0}^{\infty} w_k$ is said to converge conditionally if it converges but does not converge absolutely.

Example 2.1   The series $\displaystyle\sum_{k=0}^{\infty} \frac{i^k}{(k+1)}$ converges conditionally. (The real and imaginary parts are convergent, by the alternating series test.)

Proposition 2.4   *Every absolutely convergent series is convergent.*

Proof.   Suppose that $\sum_{k=0}^{\infty} w_k$ is absolutely convergent. Then $\sum_{k=0}^{\infty} |w_k|$ is convergent and its partial sums $T_n = \sum_{k=0}^{n} |w_k|$ form a Cauchy sequence. The generalized triangle inequality $|\sum_{k=m+1}^{n} w_k| \leq \sum_{k=m+1}^{n} |w_k|$ implies that the partial sums $S_n = \sum_{k=0}^{n} w_k$ form a Cauchy sequence in $\mathbb{C}$ and therefore converge.   □

It turns out that absolutely convergent (power) series play a central rôle in complex analysis, so any indications as to whether or not a given series is absolutely convergent are welcome. The nth-root test is one such. To discuss this, we recall some terminology. For the moment we consider only real sequences.

Let (α_n) be a sequence of real numbers which is bounded from above; that is, there is some $M \in \mathbb{R}$ such that $\alpha_n \leq M$ for all n. For each $k \in \mathbb{N}$, set $\beta_k = \sup_{n \geq k} \alpha_n$. Since $\alpha_n \leq M$, it follows that $A_k = \{\alpha_n : n \geq k\}$ is a bounded set of real numbers and so β_k is well-defined. Furthermore, since

M is an upper bound for A_k, it follows that $\beta_k \leq M$. Next, we observe that $A_{k+1} \subseteq A_k$ and so any upper bound for A_k is certainly an upper bound for A_{k+1}. Hence β_k is an upper bound for A_{k+1} and therefore $\beta_{k+1} \leq \beta_k$, since $\beta_{k+1} = \sup A_{k+1}$. We see that $(\beta_k)_{k \in \mathbb{N}}$ is a decreasing (i.e., non-increasing) sequence of real numbers. It follows that either (β_k) is bounded from below, in which case there is some $\beta \in \mathbb{R}$ such that $\beta_k \to \beta$ as $n \to \infty$ (in fact, $\beta = \inf_k \beta_k$) or else (β_k) is not bounded from below and so neither is (α_n).

In the first case, we call β the limit superior of the sequence (α_n) and write

$$\beta = \limsup_{n \to \infty} \alpha_n \quad \text{or, alternatively,} \quad \beta = \overline{\lim_n} \, \alpha_n.$$

If (β_k) is not bounded from below, we write $\limsup \alpha_n = -\infty$. Similarly, if (α_n) is not bounded from above (so that no β_k is defined), then we indicate this by writing $\limsup \alpha_n = \infty$. Note that these last two "equalities" are no such thing, but simply convenient and suggestive shorthand pieces of notation.

Examples 2.2

(1) If $\alpha_n = \frac{1}{n}$ then $\limsup_n \alpha_n = 0$.
(2) If $\alpha_n = (-1)^n$ then $\limsup_n \alpha_n = 1$.
(3) If $\alpha_n = 2^n$ then $\limsup_n \alpha_n = \infty$ (i.e., (α_n) is not bounded from above).
(4) If $\alpha_n = -n + (-1)^n n$ then $\limsup_n \alpha_n = 0$ (but notice that (α_n) is not bounded from below).

Remark 2.4 The value of $\limsup_n \alpha_n$ is unchanged by the alteration of the values of any finite number of the α_ns. Indeed, suppose that (α'_n) is another sequence in $\mathbb{R}$ such that $\alpha'_n = \alpha_n$ for all $n > N$. Then $\beta'_k \equiv \sup_{n \geq k} \alpha'_n = \sup_{n \geq k} \alpha_n = \beta_k$ for all $k > N$ and so the sequences (β'_k) and (β_k) either both converge to the same limit or are both unbounded from below.

2.7 n^{th}-Root Test

We can now discuss the so-called n^{th}-Root Test for absolute convergence of a complex series.

Theorem 2.4 (n^{th}-Root Test) *Let $(w_n)_{n \geq 0}$ be a given sequence of complex numbers and suppose that $\limsup_{n \in \mathbb{N}} |w_n|^{1/n} = L$. Then the series*

$\sum_{n=0}^{\infty} w_n$ *converges absolutely if $L < 1$ and diverges if $L > 1$.*

Proof. For any $k \in \mathbb{N}$, let $\beta_k = \sup_{n \geq k} |w_n|^{1/n}$.

Suppose first that $L < 1$ and let $r \in \mathbb{R}$ be such that $L < r < 1$. Since $\beta_k \to L$, it follows that for all sufficiently large k, we have $r > \beta_k \geq L$. In other words, there is $K \in \mathbb{N}$ such that $|w_n|^{1/n} < r$ whenever $n > K$. Taking n^{th} powers, we deduce that $|w_n| < r^n$ whenever $n > K$. It follows that the series $\sum_{n=0}^{\infty} |w_n|$ converges by comparison with the convergent geometric series $\sum_{n=0}^{\infty} r^n$.

Suppose now that $L > 1$. Then $\beta_k \geq L > 1$ for all k. It follows from the definition of the supremum that for any given $\delta > 0$, there is some $m > k$ such that $|w_m|^{1/m} > \sup_{n \geq k} |w_n|^{1/n} - \delta$. In particular, if we take δ to be $L - 1$, then it follows that, for any k, there is some $m > k$ such that

$$|w_m|^{1/m} > \beta_k - \delta > L - \delta = L - (L-1) = 1.$$

But then $|w_m| \geq 1$ for such m and it is false that $|w_n| \to 0$ as $n \to \infty$. We conclude that $\sum_{n=0}^{\infty} w_n$ cannot be convergent. $\square$

Remark 2.5 Note that the theorem says nothing about the situation when $L = 1$. This is because the information that $L = 1$ is simply insufficient to determine the convergence or otherwise of the series. For example, if $w_0 = 1$ and $w_n = 1/n$, for $n \geq 1$, then $|w_n|^{1/n} = 1/n^{1/n} \to 1$ as $n \to \infty$ and $L = 1$. The series $\sum_{n=1}^{\infty} 1/n$ is not convergent. However, putting $v_n = w_n^2$, we still have $\limsup |v_n|^{1/n} = 1$ but now $\sum_{n=0}^{\infty} v_n$ is absolutely convergent.

2.8 Ratio Test

We can recover the Ratio Test as a corollary.

Theorem 2.5 (Ratio Test) *Suppose that $w_n \neq 0$, for all $n = 0, 1, \ldots$ and that the ratios $|w_{n+1}|/|w_n| \to L$, as $n \to \infty$. Then if $L < 1$ the series $\sum_{n=0}^{\infty} w_n$ converges absolutely, and if $L > 1$, it diverges.*

Proof. To simplify the discussion, we introduce a harmless adjustment to the series $\sum_{n=0}^{\infty} w_n$. Indeed, if $\kappa \neq 0$ and $v_n = \kappa w_n$, then $\sum_{n=0}^{\infty} w_n$ converges absolutely (or diverges) if and only if the same is true of the series $\sum_{n=0}^{\infty} v_n$.

For any given $\varepsilon > 0$, there is some $N \in \mathbb{N}$ such that

$$L - \varepsilon < \frac{|w_{n+1}|}{|w_n|} < L + \varepsilon \qquad (*)$$

whenever $n > N$. For given μ, let $\kappa = \mu^{N+1} / |w_{N+1}|$ so $|v_{N+1}| = \mu^{N+1}$. Then, for $n > N + 1$,

$$|v_n| = \frac{|v_n|}{|v_{n-1}|} \frac{|v_{n-1}|}{|v_{n-2}|} \cdots \frac{|v_{N+2}|}{|v_{N+1}|} \frac{|v_{N+1}|}{\mu^{N+1}} \mu^{N+1}$$

$$= \frac{|w_n|}{|w_{n-1}|} \frac{|w_{n-1}|}{|w_{n-2}|} \cdots \frac{|w_{N+2}|}{|w_{N+1}|} \mu^{N+1}. \qquad (**)$$

Suppose $L < 1$. Then we may choose $L < \mu < 1$ and we may suppose that $\varepsilon > 0$ is so small that $L + \varepsilon < \mu$. From $(*)$ and $(**)$, we see that

$$|v_n| < \mu^n$$

for all $n > N + 1$ and therefore $\limsup_n |v_n|^{1/n} \le \mu < 1$. Hence the series $\sum_{n=0}^{\infty} v_n$ and consequently $\sum_{n=0}^{\infty} w_n$ converges absolutely, by the nth-root test.

On the other hand, if $L > 1$, then we may choose $1 < \mu < L$ and we may suppose that $\varepsilon > 0$ is so small that $\mu < L - \varepsilon$. But then $(*)$ and $(**)$ imply that

$$|v_n| > \mu^n$$

for all $n > N + 1$ and so $\limsup_n |v_n|^{1/n} \ge \mu > 1$ and we conclude that $\sum_{n=0}^{\infty} v_n$ and therefore $\sum_{n=0}^{\infty} w_n$ diverges. $\qquad \square$

As before, the series with $w_n = 1/n$ for $n \ge 1$ (and $w_0 = 1$, say) or with $w_n = 1/n^2$ for $n \ge 1$ (and $w_0 = 1$) show that nothing can be said when $L = 1$.

Chapter 3

Metric Space Properties of the Complex Plane

3.1 Open Discs and Interior Points

In this chapter we introduce some terminology and discuss some properties of various types of subsets of $\mathbb{C}$. The ideas readily extend to the more general setting of metric spaces.

Definition 3.1 The open disc with centre z_0 and radius $r > 0$ is the set
$$D(z_0, r) \equiv \{ z \in \mathbb{C} : |z - z_0| < r \}.$$

Notice that the circumference of the disc, the set $\{ z : |z - z_0| = r \}$, is not included in the set $D(z_0, r)$.

Intuitively, we might think of a typical set in $\mathbb{C}$ as having an "edge" or boundary together with an "inside" or interior. An interior point of a given set should be completely surrounded by points of the set. The following definition captures this idea.

Definition 3.2 A point w in a given set S is said to be an interior point of S if there is some $r > 0$ such that $D(w, r) \subseteq S$.

Thus, w is an interior point of the set S if and only if it is the centre of some open disc made up entirely of points from S itself.

Examples 3.1

(1) Let $S = \{ z : \operatorname{Im} z > 0 \}$. Every point $w \in S$ is an interior point; indeed, for any $w \in S$, $D(w, r) \subseteq S$ whenever $0 < r < \operatorname{Im} w$. (If $\zeta \in D(w, r)$, then $\operatorname{Im} \zeta = \operatorname{Im} w + \operatorname{Im}(\zeta - w) \geq \operatorname{Im} w - |\zeta - w| > \operatorname{Im} w - r > 0$.)
(2) Let $S = \{ z : \operatorname{Im} z \geq 0 \}$. Then w is an interior point of S whenever $\operatorname{Im} w > 0$, as above. If $\operatorname{Im} w = 0$, then w is *not* an interior point of S, since $D(w, r)$ will always contain points z with $\operatorname{Im} z < 0$ no matter

what $r > 0$ is chosen (for example, the point $w - ir/2$) and so cannot be contained in S.

(3) The set $\{\, z : \operatorname{Re} z = 0 \,\}$ contains no interior points. (It is a line, and a line cannot contain discs.) More formally, let $w \in S = \{\, z : \operatorname{Re} z = 0 \,\}$. Then, for any $r > 0$, the disc $D(w, r)$ contains the point $w + ir/2$ which does not belong to S, so there is no disc, $D(w, r)$, wholly contained in S. Therefore w is not an interior point of S.

(4) Every point in $\mathbb{C}$ is an interior point of $\mathbb{C}$. (For any given point $w \in \mathbb{C}$, any $r > 0$ of your choice will be such that $D(w, r) \subseteq \mathbb{C}$. So w is an interior point of $\mathbb{C}$.)

Definition 3.3   A set $G \subseteq \mathbb{C}$ is called open if for any point $w \in G$ there is some $r > 0$ such that $D(w, r) \subseteq G$. In other words, G is open if and only if it consists entirely of interior points.

Examples 3.2

(1) $\mathbb{C}$ is open.
(2) $G = \{\, z : \operatorname{Im} z > 0 \,\}$ is open.
(3) $A = \{\, z : \operatorname{Re} z \geq 0 \,\}$ is not open. (For example, $w = i \in A$, but it is impossible to find $r > 0$ such that $D(i, r) \subseteq A$. Indeed, the disc $D(i, r)$ contains the point $-r/2 + i \notin A$.)
(4) $\varnothing$ is open. This is a rather peculiar situation. If $\varnothing$ were *not* open, then there would be some point $w \in \varnothing$ such that, for any $r > 0$, the disc $D(w, r)$ contained some point not in $\varnothing$. Since there is no such point w, we conclude that $\varnothing$ cannot fail to be open, i.e., it is open essentially by default. Whilst this result may appear a little bizarre, it is nonetheless very convenient, since it means that the intersection of two open sets is also open, even if they have no points in common—see below.

First let us show that the so-called open discs $D(z, r)$ really are open.

Proposition 3.1   *For any $z \in \mathbb{C}$ and any $r > 0$, the set $D(z, r)$ is open.*

Proof.   Let $w \in D(z, r)$ be given. Then $|w - z| < r$. Let $\rho > 0$ be such that $0 < \rho < r - |w - z|$ (note that the right hand side is positive). Then $|w - z| < r - \rho$. We claim that the disc $D(w, \rho)$ is contained in $D(z, r)$.

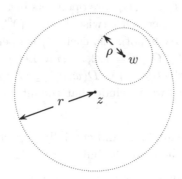

Fig. 3.1   The set $D(z, r)$ is open.

To see this, let $\zeta \in D(w, \rho)$. Then we have

$$
\begin{aligned}
|\zeta - z| &= |(\zeta - w) + (w - z)| \\
&\leq |\zeta - w| + |w - z|, \text{ by the triangle inequality,} \\
&< \rho + (r - \rho) \\
&= r,
\end{aligned}
$$

as required. Since $w \in D(z, r)$ is arbitrary, we conclude that $D(z, r)$ is open. ☐

Remark 3.1   This proposition justifies the terminology "open disc". The set $\{ z : |z - w| \leq r \}$ consists of the "inside" (interior) together with the circumference (boundary) of the disc, and is called the closed disc, denoted by $\overline{D(w, r)}$. We shall discuss closed sets shortly, as well as this "overbar" notation.

Theorem 3.1

(i) *Let $\{G_\alpha\}_{\alpha \in I}$ be an arbitrary collection of open sets in $\mathbb{C}$, indexed by the set I. Then the union $\bigcup_{\alpha \in I} G_\alpha$ is an open set.*

(ii) *Let $G_1, \ldots, G_m$ be any finite collection of open sets in $\mathbb{C}$. Then their intersection $G_1 \cap G_2 \cap \cdots \cap G_m$ is an open set.*

Proof.   Let $\{G_\alpha\}$ be given. If they are all empty, then so is their union and we know that $\varnothing$ is open. Otherwise, let $\zeta \in \bigcup_\alpha G_\alpha$. Then there is some $\alpha_0 \in I$ such that $\zeta \in G_{\alpha_0}$. By hypothesis, G_{α_0} is open and so there is some $r > 0$ such that $D(\zeta, r) \subseteq G_{\alpha_0}$. But then it follows that $D(\zeta, r) \subseteq \bigcup_\alpha G_\alpha$ and so $\bigcup_\alpha G_\alpha$ is open.

Suppose now that $G_1, \ldots, G_m$ are open sets in $\mathbb{C}$. If $\bigcap_{j=1}^{m} G_j = \varnothing$, then we are done, since $\varnothing$ is open. Otherwise, let $w \in \bigcap_{j=1}^{m} G_j$. Then $w \in G_j$ for each $j = 1, 2, \ldots, m$. By hypothesis, each G_j is open and so there is some $r_j > 0$ such $D(w, r_j) \subseteq G_j$, $j = 1, 2, \ldots, m$. Let $r = \min\{r_1, r_2, \ldots, r_m\}$. Then $r > 0$ and clearly $D(w, r) \subseteq D(w, r_j) \subseteq G_j$, $j = 1, 2, \ldots, m$. Hence $D(w, r) \subseteq \bigcap_{j=1}^{m} G_j$ and we conclude that the intersection $G_1 \cap \cdots \cap G_m$ is open. $\qquad\square$

Remark 3.2　　These are basic and crucially important properties of open sets. Abstractions of these are the building blocks for a topological space.

Remark 3.3　　Note that it is false in general that an arbitrary intersection of open sets is open. For example, for each $n \in \mathbb{N}$, let G_n be the open disc $D(0, 1/n) = \{ z : |z| < 1/n \}$. Then each G_n is open, but their intersection is $\bigcap_{n=1}^{\infty} G_n = \{ z : z \in G_n \text{ for every } n \} = \{ z : |z| < 1/n \text{ all } n \in \mathbb{N} \} = \{0\}$ which is not an open set (there is no $r > 0$ such that $D(0, r) \subseteq \{0\}$).

Remark 3.4　　We can express convergence of sequences using discs. The sequence (z_n) converges to z in $\mathbb{C}$ provided that for any given $\varepsilon > 0$ it is true that z_n is eventually within ε of z, i.e., there is some $N \in \mathbb{N}$ such that $n > N$ implies that $|z_n - z| < \varepsilon$. This is the same as saying that $z_n \in D(z, \varepsilon)$, whenever $n > N$. Thus, $z_n \to z$ if and only if (z_n) is eventually in any given disc $D(z, \varepsilon)$, $\varepsilon > 0$ (centred on z). The mental image of a convergent sequence in $\mathbb{C}$ is one of a collection of dots (representing the points of the sequence) eventually moving inside (and staying inside) any given disc around the limit.

3.2　Closed Sets

Next, we define closed sets.

Definition 3.4　　A set F in $\mathbb{C}$ is closed if its complement $\mathbb{C} \setminus F$ is open. In other words, closed sets are precisely the complements of open sets, and vice versa.

Examples 3.3

(1) The empty set $\varnothing$ is closed, because its complement, $\mathbb{C}$, is open. The whole complex plane, $\mathbb{C}$, is closed, because its complement, $\varnothing$, is open.
(2) The set $F = \{ z : \operatorname{Im} z \geq 0 \}$ is closed. To see this, we just have to show that its complement is open. But if $w \notin F$, then we must have

Im $w < 0$. Putting $r = \frac{1}{2} |\text{Im } w|$ we see that $D(w, r) \subseteq \mathbb{C} \setminus F$ and we deduce that $\mathbb{C} \setminus F$ is open, i.e., F is closed.

The following is a very useful characterization of closed sets.

Proposition 3.2 *The non-empty set F in $\mathbb{C}$ is closed if and only if every sequence in the set F which converges in $\mathbb{C}$ has its limit in F; i.e., F is closed if and only if whenever (z_n) is a sequence in F such that $z_n \to z$, for some $z \in \mathbb{C}$, as $n \to \infty$, then it is true that $z \in F$.*

Proof. Suppose that F is closed and suppose that (a_n) is a given sequence in F such that $a_n \to z$, as $n \to \infty$. We must show that $z \in F$. Suppose, on the contrary, that $z \notin F$. Then z belongs to the open set $\mathbb{C} \setminus F$. Hence there is some $r > 0$ such that $D(z, r) \subseteq \mathbb{C} \setminus F$. In particular, this means that $D(z, r) \cap F = \varnothing$. However, $a_n \to z$ means that $a_n \in D(z, r)$ for all sufficiently large n. This is a contradiction and we conclude that $z \in F$, as required.

For the converse, suppose that $a_n \to z$ with $a_n \in F$ implies that $z \in F$. We must show that $\mathbb{C} \setminus F$ is open. If $F = \mathbb{C}$, there is nothing to prove, so suppose that $F \neq \mathbb{C}$. Let $z \in \mathbb{C} \setminus F$. We claim that there is $r > 0$ such that $D(z, r) \subseteq \mathbb{C} \setminus F$. If this were not true, then, no matter what choice of $r > 0$ we make, we will have $D(z, r) \cap F \neq \varnothing$. In particular, for each $n \in \mathbb{N}$, we would have $D(z, \frac{1}{n}) \cap F \neq \varnothing$. So suppose $a_n \in D(z, \frac{1}{n}) \cap F$. Then $a_n \in F$ and $|a_n - z| < \frac{1}{n}$, for all $n \in \mathbb{N}$. It follows that $a_n \to z$, as $n \to \infty$. By hypothesis, this means that $z \in F$, a contradiction.

We conclude that there is indeed some $r > 0$ such that $D(z, r) \subseteq \mathbb{C} \setminus F$. Therefore $\mathbb{C} \setminus F$ is open and so F is closed. □

Remark 3.5 So the set F fails to be closed if there exists some sequence of points in F which converges to a point *not* belonging to F.

In fact, we could therefore *define* a set F to be closed if (a_n) in F and $a_n \to z$ in $\mathbb{C}$ implies that $z \in F$.

Example 3.4 The disc $D(w, r)$ is not closed. To see this, we can take, for example, a sequence of points converging radially outwards towards a point on the circumference of the disc. Specifically, consider the sequence $z_n = w + (1 - \frac{1}{n})r$, $n \in \mathbb{N}$. Then $z_n \in D(w, r)$, for each n, but $z_n \to w + r$ which does not belong to $D(w, r)$.

Remark 3.6 A given set $A \subseteq \mathbb{C}$ need be neither open nor closed. For example, the set $S = \{ z : 0 < \text{Re } z \leq 1 \}$ is not open, nor is it closed. Indeed, $1 \in S$, but there is no $r > 0$ such that $D(1, r) \subseteq S$, so 1 is not an interior point, and therefore S is not open. On the other hand, the sequence $z_n = \frac{1}{n}$, $n \in \mathbb{N}$, belongs to S and converges to $0 \notin S$. So S is not closed.

So there are sets which are neither open nor closed, but is it possible to have sets which are *both* open *and* closed? We have seen that $\mathbb{C}$ and $\varnothing$ have this property. It is no use trying to find further examples: we will see shortly that these are the only possibilities.

3.3   Limit Points

Closed sets can also be characterized in terms of limit points, which we now consider.

Definition 3.5   The point w is said to be a limit point of a given subset A of $\mathbb{C}$ if *every* disc $D(w, r)$, $r > 0$, contains some point of A other than w. Limit points are also called cluster points or accumulation points.

Note that the point w may or may not actually belong to A. If $D'(w, r)$ denotes the punctured disc $\{ z : 0 < |z - w| < r \}$, then the definition says that w is a limit point of A if $D'(w, r) \cap A \neq \varnothing$ for all $r > 0$. The idea is that there should be points of A not equal to, but arbitrarily close to w.

By setting $r = \frac{1}{n}$ for each n in $\mathbb{N}$ we see that w is a limit point of A if and only if there is some sequence (a_n) of points in A, all *different* from w, such that $a_n \to w$, as $n \to \infty$.

Examples 3.5

(1) If G is a non-empty open set, then every point of G is a limit point of G. Indeed, if $w \in G$, then there is some $\rho > 0$ such that $D(w, \rho) \subseteq G$. Hence, for any $r > 0$,

$$D'(w, r) \cap G \supseteq D'(w, R) \neq \varnothing$$

where $R = \min\{r, \rho\}$.

(2) The point $w = i$ is a limit point of the unit disc $D(0, 1)$. To see this, let $z_t = i - it = (1 - t)i$ for $0 < t < 1$. Then $z_t \in D(0, 1)$ and $|z_t - i| = t$. It follows that $z_t \neq i$ and $z_t \in D'(i, r)$ whenever $0 < t < r$. Hence i is a limit point of the unit disc. In fact, every point on the circumference of the disc is a limit point of the disc (take a sequence inside the disc which converges radially outwards to the point of the circumference). The set of limit points of the disc $D(\zeta, r)$ is the set $\{ z : |z - \zeta| \leq r \}$.

(3) If A contains a finite number of points, then it has no limit points. This is clear if $A = \{\, a \,\}$, a singleton set. Otherwise, for any $w \in \mathbb{C}$, let $\rho = \min\{\, |w - a| : a \in A \text{ and } a \neq w \,\}$. Then $\rho > 0$ and the punctured disc $D'(w, \rho)$ contains no points of A, so w cannot be a limit point of A.

Proposition 3.3 *The set F is closed if and only if F contains all its limit points.*

Proof. Suppose first that F is closed, i.e., F^c, the complement of F, is open. Suppose that ζ is a limit point of F. We must show that $\zeta \in F$. If this were not the case, then we would have $\zeta \in F^c$. But then this would mean that there was some $r > 0$ such that $D(\zeta, r) \subseteq F^c$ (because F^c is open, by hypothesis). In particular, $D(\zeta, r) \cap F = \varnothing$. But for ζ to be a limit point of F, we must have that $D'(\zeta, r) \cap F \neq \varnothing$, which is a contradiction. We conclude that $\zeta \in F$, as required.

Next, suppose that F contains all its limit points. We must show that F^c is open. If $F^c = \varnothing$, then we are done. So suppose that $F^c \neq \varnothing$ and let $w \in F^c$. We wish to prove that there is some $r > 0$ such that $D(w, r) \subseteq F^c$. If this were not true, then no disc $D(w, r)$ would be wholly contained in F^c. This is just the statement that $D(w, r) \cap F \neq \varnothing$ for every $r > 0$. But $w \notin F$, by hypothesis, and so we may say that $D'(w, r) \neq \varnothing$ for all $r > 0$. This means that w is a limit point of F and so must belong to F, if F is assumed to contain all its limit points. We have a contradiction. We conclude that, indeed, F^c is open and so F is closed. $\qquad\square$

Remark 3.7 The statement that a set A contains all its limit points is equivalent to the statement that whenever (a_n) is a convergent sequence in A then its limit is also in A.

To see this, we first note that if w is a limit point of A, then it is the limit of some sequence (a_n) in A and so the first statement above follows from the second. Suppose now that (a_n) is a sequence in A and that $a_n \to w$. If $w = a_k$ for some k, then $w \in A$ (because $a_k \in A$). On the other hand, if $a_n \neq w$ for all n, then this means that w is a limit point of A. Hence the second statement follows from the first.

We therefore have three equivalent statements, namely:

- the complement of F is open;
- F contains the limit of any of its convergent sequences (that is, if (z_n) belongs to F and $z_n \to z$ for $z \in \mathbb{C}$, then z actually also belongs to F);
- F contains all its limit points.

What's going on ?   The equivalence of these statements means that a closed set in $\mathbb{C}$ can be defined (equivalently) in several ways. Which formulation one chooses amounts largely to personal preference. To each definition adopted, one then has two propositions of the form "F is closed if and only if ...", where "..." denotes either of the other two equivalent statements. It turns out that the most appropriate definition is the one we have given, namely that in terms of the complement being open. This generalizes to metric spaces and also to the theory of topological spaces, where the use of sequences is generally not the best approach to take.

3.4   Closure of a Set

Definition 3.6   For any $A \subseteq \mathbb{C}$, the closure of A is the set

$$\overline{A} = \{\, z : \text{there is some sequence } (a_n) \text{ in } A \text{ such that } a_n \to z \,\}.$$

In words, $\overline{A}$ consists of all those points of $\mathbb{C}$ which are the limits of some sequence from A. In particular, for any $a \in A$, we can set $a_n = a$, for all $n \in \mathbb{N}$. Then, trivially, $a_n \to a$ and so $a \in \overline{A}$ and therefore $A \subseteq \overline{A}$. Furthermore, if A is closed, then (a_n) in A and $a_n \to z$ implies that z belongs to A. Therefore $\overline{A} = A$ if A is closed.

Remark 3.8   Alternatively, one could define the closure of a set A to be the set A itself together with all its limit points. This follows because any point $w \notin A$ is a limit point of A if and only if w is the limit of a sequence of points of A.

Example 3.6   What is the closure, K, say, of the disc $D(w,r)$? We know that $D(w,r) \subseteq K$. Let $\zeta \in K$. Then, by definition, there is some sequence (z_n) in $D(w,r)$ such that $z_n \to \zeta$, as $n \to \infty$. In particular, $|z_n - w| < r$ for each $n \in \mathbb{N}$. But $z_n \to \zeta$ implies that $z_n - w \to \zeta - w$ and therefore $|z_n - w| \to |\zeta - w|$, as $n \to \infty$. Since $|z_n - w| < r$ for each $n \in \mathbb{N}$, it follows that $|\zeta - w| \le r$ (the real sequence $(|z_n - w|)$ lies in the interval $[0,r]$, so, therefore, must its limit). We deduce that

$$\{\, z : |z - w| < r \,\} \subseteq K \subseteq \{\, z : |z - w| \le r \,\}.$$

Now let z be any point in $\mathbb{C}$ with $|z - w| = r$. We shall show that $z \in K$. Indeed, let (z_n) be the sequence of points "moving along the radius of $D(w,r)$ from w towards z" given by $z_n = w + (1 - \frac{1}{n})(z - w)$. We have $|z_n - w| = |(1 - \frac{1}{n})(z - w)| = (1 - \frac{1}{n})|z - w| < r$, so that $z_n \in D(w,r)$, for each n. Clearly $z_n \to w + (z - w) = z$ as $n \to \infty$ and we conclude that $z \in K$, as claimed.

It follows that $K = \{ z : |z - w| \le r \}$; which justifies our notation for this set, namely, $\overline{D(w, r)}$, as given earlier (we shall also shortly justify the terminology "closed disc").

Proposition 3.4 *For any set A, its closure $\overline{A}$ is a closed set.*

Proof. If $\overline{A} = \varnothing$, we are done. So suppose that (z_n) is a sequence in $\overline{A}$ and that $z_n \to \zeta$ as $n \to \infty$. We must show that $\zeta \in \overline{A}$. By definition of $\overline{A}$, this follows if we can show that there is some sequence (a_n) in A with $a_n \to \zeta$.

To see this, we note that for each n, $z_n \in \overline{A}$, which means that z_n is itself the limit of some sequence of points from A. In particular, any such sequence is eventually within $1/n$ of z_n, that is, there is certainly some element of A within distance $1/n$ of z_n. Let a_n be one such element of A; $a_n \in A$ and $|z_n - a_n| < 1/n$. Doing this for each n yields a sequence (a_n) of elements of A. We claim that $a_n \to \zeta$, as $n \to \infty$. Let $\varepsilon > 0$ be given. Then there is $N_0 \in \mathbb{N}$ such that $n > N_0$ implies that $|z_n - \zeta| < \varepsilon/2$. Let $N = \max\{N_0, 2/\varepsilon\}$, so that, in particular, $1/n < \varepsilon/2$ whenever $n > N$. We have

$$|a_n - \zeta| \le |a_n - z_n| + |z_n - \zeta|$$
$$< \tfrac{1}{n} + \tfrac{1}{2}\varepsilon, \quad \text{whenever } n > N_0,$$
$$< \tfrac{1}{2}\varepsilon + \tfrac{1}{2}\varepsilon = \varepsilon$$

whenever $n > N$. It follows that $a_n \to \zeta$ giving $\zeta \in \overline{A}$ and we conclude that $\overline{A}$ is closed. $\qquad\square$

Remark 3.9 This result justifies the terminology "closed disc" for the set $\overline{D(w, r)}$.

Remark 3.10 The above result implies that the set A together with its limit points is closed. This means that this "enlarged" set has no new limit points. Any limit point of $\overline{A}$ is already a limit point of A. This can be seen directly as follows. Let ζ be a limit point of $\overline{A}$ and let $r > 0$ be given. Then there is some $w \in \overline{A}$ with $w \ne \zeta$ and $|\zeta - w| < \tfrac{1}{2}r$. Now, $w \in \overline{A}$, and so for any given $\rho > 0$ there is some $a \in A$ with $a \ne w$ such that $|w - a| < \rho$. In particular, we may take $\rho = |\zeta - w|$ so that $a \ne \zeta$ (it is closer to w than ζ is) so that $|\zeta - a| \le |\zeta - w| + |w - a| < r$ and it follows that ζ is a limit point of A.

Since a set is closed if and only if it contains its limit points, we see that the above discussion constitutes, in fact, an alternative proof that the closure of any set is closed.

3.5 Boundary of a Set

Next, we define the boundary of a set. It is usually quite clear from a diagram what the boundary of a set is, but we still need a formal definition.

Definition 3.7   The boundary of a set $A \subseteq \mathbb{C}$ is the set

$$\partial A \equiv \{\, z : \text{for all } r > 0,\ D(z,r) \cap A \neq \varnothing \text{ and } D(z,r) \cap (\mathbb{C} \setminus A) \neq \varnothing \,\}.$$

In other words, a point z belongs to the boundary of the set A provided *every* open disc around z contains both points of A and points not in A.

The next result is a useful characterization of the boundary of a set.

Proposition 3.5   *For given $A \subseteq \mathbb{C}$, the point z belongs to ∂A if and only if there is some sequence (a_n) in A and some sequence (b_n) in $\mathbb{C} \setminus A$ such that $a_n \to z$ and $b_n \to z$, as $n \to \infty$.*

Proof.   First suppose that $z \in \partial A$. Then, for any $n \in \mathbb{N}$, the sets $D(z, \frac{1}{n}) \cap A$ and $D(z, \frac{1}{n}) \cap (\mathbb{C} \setminus A)$ are non-empty. Let a_n be any point of $D(z, \frac{1}{n}) \cap A$ and let b_n be any point of $D(z, \frac{1}{n}) \cap (\mathbb{C} \setminus A)$. In this way, we have constructed sequences (a_n) in A and (b_n) in $\mathbb{C} \setminus A$. Furthermore, $|z - a_n| < \frac{1}{n}$ since $a_n \in D(z, \frac{1}{n})$, and so $a_n \to z$ as $n \to \infty$. Similarly, $|z - b_n| < \frac{1}{n}$, and so $b_n \to z$ as $n \to \infty$.

Now suppose that there are sequences (a_n), (b_n) such that $a_n \in A$, $b_n \notin A$, for all n and $a_n \to z$ and $b_n \to z$, as $n \to \infty$. Let $r > 0$ be given. We must show that both $D(z,r) \cap A$ and $D(z,r) \cap (\mathbb{C} \setminus A)$ are non-empty. But a_n converges to z and so (a_n) is eventually in $D(z,r)$. This implies that $D(z,r) \cap A \neq \varnothing$. Similarly, $b_n \to z$ and therefore (b_n) is eventually in $D(z,r)$ and so $D(z,r) \cap (\mathbb{C} \setminus A) \neq \varnothing$. $\square$

From this result, we deduce that

$$\partial A = \overline{A} \cap \overline{\mathbb{C} \setminus A}.$$

Example 3.7   The sets $D(w,r)$ and $\overline{D(w,r)}$ have the same boundary,

$$\partial D(w,r) = \partial \overline{D(w,r)} = \{\, z : |z - w| = r \,\}.$$

Example 3.8   Let A be the horizontal strip $A = \{ z : -1 < \operatorname{Im} z < 1 \}$. It is clear from a diagram that the boundary is made up of the two lines $L_1 = \{ z : \operatorname{Im} z = -1 \}$ and $L_2 = \{ z : \operatorname{Im} z = 1 \}$. This can be seen directly from the definition, but it also follows quite easily from the proposition above. Indeed, if $\zeta = \alpha - i \in L_1$, then setting $a_n = \alpha - (1 - \frac{1}{n})i$ and $b_n = \alpha - (1 + \frac{1}{n})i$, we see that $a_n \in A$, $b_n \notin A$ and both a_n and b_n converge to ζ. Hence $\zeta \in \partial A$. A similar argument shows that any point of L_2 belongs to ∂A. (In fact, if $w = \alpha + i \in L_2$, then $\overline{a_n} \in A$, $\overline{b_n} \notin A$ and both $\overline{a_n}$ and $\overline{b_n}$ converge to w.) If $|\operatorname{Im} z| < 1$, then clearly no sequence from A^c can converge to z. (This is because if $z_n \to z$, then $\operatorname{Im} z_n \to \operatorname{Im} z$. But $z_n \notin A$ implies that $|\operatorname{Im} z_n| \geq 1$ and the imaginary part of any limit of z_n would also have to satisfy this inequality.) Similarly, if $|\operatorname{Im} z| > 1$, then no sequence from A can converge to z.

The following is the counterpart of theorem 3.1 for closed sets—but notice that now the union must be finite whereas it is the intersection which can be arbitrary.

Theorem 3.2

(i) *For any finite collection* $F_1, \ldots, F_m$ *of closed subsets of* $\mathbb{C}$, *the union* $F_1 \cup F_2 \cup \cdots \cup F_m$ *is a closed set.*

(ii) *Let* $\{F_\alpha\}_{\alpha \in I}$ *be an arbitrary collection of closed sets in* $\mathbb{C}$, *indexed by the set* I. *Then the intersection* $\bigcap_{\alpha \in I} F_\alpha$ *is a closed set.*

Proof.   This follows from theorem 3.1 by taking complements. Indeed, if $F_1, \ldots, F_m$ are closed, then the sets $\mathbb{C} \setminus F_1, \ldots, \mathbb{C} \setminus F_m$ are open. Hence, by theorem 3.1,

$$\mathbb{C} \setminus (F_1 \cup \cdots \cup F_m) = (\mathbb{C} \setminus F_1) \cap \cdots \cap (\mathbb{C} \setminus F_m)$$

is open and we deduce that $F_1 \cup \cdots \cup F_m$ is closed.

For any collection $\{F_\alpha\}$ of subsets of $\mathbb{C}$, we have

$$\mathbb{C} \setminus \left(\bigcap_\alpha F_\alpha \right) = \bigcup_\alpha \mathbb{C} \setminus F_\alpha.$$

Now, if each F_α is closed, each $(\mathbb{C} \setminus F_\alpha)$ is open and so, by theorem 3.1, $\bigcup_\alpha \mathbb{C} \setminus F_\alpha$ is open. Hence $\mathbb{C} \setminus (\bigcap_\alpha F_\alpha)$ is open and so $\bigcap_\alpha F_\alpha$ is closed. $\square$

3.6 Cantor's Theorem

Example 3.9   For $n \in \mathbb{N}$, let $F_n = \{\, z : \operatorname{Re} z \geq \frac{1}{n} \,\}$. Then each F_n is closed, but $\bigcup_n F_n = \{\, z : \operatorname{Re} z > 0 \,\}$ which is not closed.

The next theorem sheds some light onto questions such as "what do you end up with if you take a set and keep cutting it in half (and throwing away the half that has been cut off)?" It will make a crucial appearance in the middle of arguably the most important theorem in the subject, namely, Cauchy's Theorem.

Definition 3.8   A sequence $(A_n)_{n \in \mathbb{N}}$ of sets is said to be nested if it is true that $A_{n+1} \subseteq A_n$, for all $n \in \mathbb{N}$.

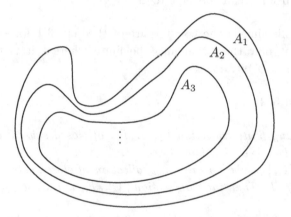

Fig. 3.2   A nested sequence of sets in $\mathbb{C}$.

Definition 3.9   A non-empty set $A \subseteq \mathbb{C}$ is said to be bounded if there is some $M > 0$ such that $|z| < M$ for all $z \in A$. If A is bounded, its diameter is defined to be the number $\operatorname{diam} A = \sup\{\, |z - \zeta| : z, \zeta \in A \,\}$.

So the set A is bounded if and only if it is contained inside some disc, $A \subseteq D(0, M)$. Notice also that if $|z| < M$ for all $z \in A$, then it is true that $|z - \zeta| \leq |z| + |\zeta| < 2M$ for all $z, \zeta \in A$ and so $\operatorname{diam} A$ is well-defined (and is not greater than $2M$).

Theorem 3.3 (Cantor's Theorem)   *Suppose that (F_n) is a nested sequence of (non-empty) bounded, closed sets such that $\operatorname{diam} F_n \to 0$ as $n \to \infty$. Then $\bigcap_{n=1}^{\infty} F_n$ consists of exactly one point.*

Proof. First we shall show that $\bigcap_n F_n$ can contain *at most* one point. To see this, suppose that $z_1, z_2 \in \bigcap_n F_n$. Then $z_1, z_2 \in F_n$, for every $n \in \mathbb{N}$, and $|z_1 - z_2| \le \operatorname{diam} F_n$. Now, $\operatorname{diam} F_n \to 0$ and so for any given $\varepsilon > 0$ there is $N \in \mathbb{N}$ such that $\operatorname{diam} F_n < \varepsilon$, whenever $n > N$. In particular, for any $n > N$,

$$|z_1 - z_2| \le \operatorname{diam} F_n < \varepsilon.$$

That is, $0 \le |z_1 - z_2| < \varepsilon$ for any given $\varepsilon > 0$, which is only possible if $|z_1 - z_2| = 0$. We must have $z_1 = z_2$ which means that $\bigcap_n F_n$ can contain at most one point.

The proof is complete if we can show that $\bigcap_n F_n$ is not empty, for then it must contain precisely one point. For each $n \in \mathbb{N}$, let z_n be some point in F_n (any point at all). We claim that the sequence (z_n) thus obtained is a Cauchy sequence. For this, let $\varepsilon > 0$ be given. Since $\operatorname{diam} F_n \to 0$, there is $N \in \mathbb{N}$ such that $\operatorname{diam} F_n < \varepsilon$ whenever $n > N$. Let $m, n > N$. Then $z_m, z_n \in F_{N+1}$, since both $F_m \subseteq F_{N+1}$ and $F_n \subseteq F_{N+1}$ (because the F_ns are nested), and therefore $|z_m - z_n| \le \operatorname{diam} F_{N+1} < \varepsilon$. Hence (z_n) is a Cauchy sequence.

Now, we know that any Cauchy sequence converges, that is, there exists ζ in $\mathbb{C}$ such that $z_n \to \zeta$ as $n \to \infty$. We shall complete the proof by showing that $\zeta \in \bigcap_n F_n$. Let $k \in \mathbb{N}$ be given. For $m = 1, 2, \ldots$ define $\zeta_m = z_{k+m}$. Then clearly $\zeta_m \to \zeta$ as $m \to \infty$. (For given $\varepsilon > 0$ there is N such that $|\zeta - z_n| < \varepsilon$ whenever $n > N$. In particular, for any $m > N$, we have that $m + k > N$ and so $|\zeta - \zeta_m| = |\zeta - z_{k+m}| < \varepsilon$.) Furthermore, $\zeta_m = z_{k+m} \in F_{k+m} \subseteq F_k$ for each m and, by hypothesis, F_k is closed. Hence $\zeta \in F_k$. This holds for any $k \in \mathbb{N}$ and so we finally deduce that $\zeta \in \bigcap_n F_n$, as required. $\qquad\square$

Example 3.10 For each $n \in \mathbb{N}$, let F_n be the "half-plane" defined by $F_n = \{ z : \operatorname{Re} z \le -n \}$. Each F_n is a closed set and evidently $F_{n+1} \subseteq F_n$. However, $\bigcap_n F_n = \varnothing$. This does not contradict Cantor's Theorem because F_n is not bounded.

3.7 Compact Sets

A further very important concept is that of compactness.

Definition 3.10 A set $K \subseteq \mathbb{C}$ is compact if every sequence in K has a convergent subsequence with limit in K. Thus, K is compact if and only if whenever (z_n) is a sequence in K, there is a subsequence $(z_{n_k})_{k \in \mathbb{N}}$, say, and some $z \in K$ such that $z_{n_k} \to z$ as $k \to \infty$.

Evidently, a set A fails to be compact if it possesses a sequence which has *no* convergent subsequence with limit in A. It follows that the empty set, $\varnothing$, is compact because it cannot fail to be compact.

Example 3.11   The set $A = \{ z : 0 < |z| \leq 1 \}$ is not compact; for example, the sequence $(z_n = \frac{i}{n})_{n \in \mathbb{N}}$ converges to 0 and so every subsequence will also converge to 0. However, $0 \notin A$.

Compactness of sets in $\mathbb{C}$ is characterized by the following very useful criterion.

Theorem 3.4   *A subset $K \subseteq \mathbb{C}$ is compact if and if K is both closed and bounded.*

Proof.   Suppose first that K is closed and bounded. Let (z_n) be any given sequence in K. Since K is bounded, there is $M > 0$ such that $|z| < M$ for all $z \in K$. In particular, $|z_n| < M$, for all n. Writing $z_n = x_n + iy_n$, it follows that $|x_n| < M$ and also $|y_n| < M$, for all n, i.e., (x_n) and (y_n) are bounded sequences of real numbers. From "real analysis", we know that every bounded sequence of real numbers has a convergent subsequence (Bolzano-Weierstrass theorem). Hence, there is some subsequence (x_{n_k}) of (x_n) and $x \in \mathbb{R}$ such that $x_{n_k} \to x$ as $k \to \infty$.

We are unable to say anything about the convergence of the subsequence (y_{n_k}) of (y_n). However, we do know that $|y_{n_k}| < M$, for all k. For notational convenience, write $u_k = x_{n_k}$ and $v_k = y_{n_k}$. Then (u_k) converges to x, as $k \to \infty$. Again by the Bolzano-Weierstrass theorem, (v_k) has a convergent subsequence; say, $(v_{k_j})_{j \in \mathbb{N}}$ converges to y, as $j \to \infty$. But then (u_{k_j}) is a subsequence of (u_k) and so also converges to x, as $j \to \infty$. It follows that $u_{k_j} + iv_{k_j} \to x + iy$ as $j \to \infty$. Furthermore, by hypothesis, K is closed and $u_{k_j} + iv_{k_j} \in K$ for each j. Hence $x + iy \in K$. Since $(u_{k_j} + iv_{k_j})_{j \in \mathbb{N}}$ is a subsequence of (z_n), we conclude that K is compact.

For the converse, suppose that K is compact. First we shall show that K is closed. Let (z_n) be any sequence in K such that $z_n \to z$. We must show that $z \in K$. By compactness, there is a convergent subsequence; $z_{n_k} \to \zeta$, with $\zeta \in K$, as $k \to \infty$. But z_{n_k} is a subsequence of the convergent sequence (z_n) and so must have the same limit, that is, $\zeta = z$ and so $z \in K$. It follows that K is closed.

To show that K is bounded, we suppose the contrary, i.e., suppose that K is not bounded. Then for any $M > 0$ there is some $z \in K$ satisfying

$|z| \geq M$. In particular, for each $n \in \mathbb{N}$, there is some z_n, say, in K such that $|z_n| \geq n$. Since K is compact, (z_n) has a convergent subsequence, (z_{n_k}), say, with $z_{n_k} \to \zeta$, as $k \to \infty$, for some $\zeta \in K$. We shall show that this is incompatible with the inequalities $|z_{n_k}| \geq n_k$. Indeed, convergence to ζ implies that certainly $|\zeta - z_{n_k}|$ is eventually smaller than 1. Hence, for all sufficiently large k,

$$k \leq n_k \leq |z_{n_k}| = |(z_{n_k} - \zeta) + \zeta|$$
$$\leq |z_{n_k} - \zeta| + |\zeta|$$
$$< 1 + |\zeta| \, ,$$

which is impossible. This contradiction shows that, in fact, K is bounded and the proof is complete. $\qquad\square$

There is yet a further characterization of compactness.

Theorem 3.5 *Suppose that K is a compact subset of $\mathbb{C}$ and suppose that $\{G_\alpha\}_{\alpha \in I}$ is some family (indexed by I) of open sets such that $K \subseteq \bigcup_{\alpha \in I} G_\alpha$. Then there is a finite set $\alpha_1, \ldots, \alpha_m \in I$ such that $K \subseteq G_{\alpha_1} \cup \cdots \cup G_{\alpha_m}$; that is, any open cover of K has a finite subcover.*

Conversely, if $K \subseteq \mathbb{C}$ has the property that every open cover of K has a finite subcover, then K is compact.

Proof. Suppose that K is compact and let $\{G_\alpha\}$ be a collection of open sets such that $K \subseteq \bigcup_\alpha G_\alpha$ (i.e., $\{G_\alpha\}$ is an open cover of K). Let us suppose that there does not exist a finite collection of the G_αs whose union contains K and we shall show that this leads to a contradiction. (This is a two-dimensional version of the Heine-Borel Theorem.)

By theorem 3.4, we know that K is bounded. Hence there is some $d > 0$ such that K is contained in the closed square S_1, centred at 0 and with side equal to d (see Fig. 3.3);

$$K \subseteq S_1 \equiv \{\, z : |\operatorname{Re} z| \leq \tfrac{d}{2} \text{ and } |\operatorname{Im} z| \leq \tfrac{d}{2} \,\}.$$

By bisecting the sides of S_1, we construct 4 similar closed squares $S_1^{(1)}$, $S_1^{(2)}$, $S_1^{(3)}$, $S_1^{(4)}$ each of side $d/2$. For example, $S_1^{(1)}$ could be the (top right hand) closed square

$$S_1^{(1)} = \{\, z : 0 \leq \operatorname{Im} z \leq \tfrac{d}{2} \text{ and } 0 \leq \operatorname{Re} z \leq \tfrac{d}{2} \,\}.$$

and $S_1^{(2)}$ could be the (top left hand) closed square

$$S_1^{(2)} = \{\, z : 0 \leq \operatorname{Im} z \leq \tfrac{d}{2} \text{ and } -\tfrac{d}{2} \leq \operatorname{Re} z \leq 0 \,\}.$$

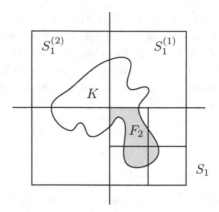

Fig. 3.3   Construct squares.

Notice that these squares always contain their boundaries, so that some (common) edges will overlap. Now, $K \subseteq S_1$ and so

$$ K = (K \cap S_1^{(1)}) \cup (K \cap S_1^{(2)}) \cup (K \cap S_1^{(3)}) \cup (K \cap S_1^{(4)}). $$

By hypothesis, it is impossible to cover K with any finite number of the G_αs and so the same must be true for at least one of the four sets $(K \cap S_1^{(i)})$, $i = 1, 2, 3, 4$. Call this set F_2 (so $F_2 = (K \cap S_1^{(i)})$ for some $i = 1, 2, 3, 4$) and, for notational convenience, let $F_1 = K$. Then $F_2 \subseteq F_1$, F_2 is a closed set, is contained in a closed square of side $d/2$ and cannot be covered by any finite collection of the G_αs. We can now divide this square into 4 similar subsquares and repeat the construction. In this way, we construct a nested sequence $F_1, F_2, F_3, \ldots$ of closed sets such that F_n is contained in a square of side $d/2^{n-1}$ (the side is halved at each step of the construction) and such that no F_n can be covered by a finite collection of the G_αs. The diameter of a square of side r is equal to $r\sqrt{2}$, and so it is clear that diam $F_n \to 0$, as $n \to \infty$. By Cantor's theorem, theorem 3.3, it follows that $\bigcap_{n=1}^{\infty} F_n = \{\zeta\}$, for some ζ. In particular, $\zeta \in F_1 = K$.

The family $\{G_\alpha\}_{\alpha \in I}$ is a cover for K and so there is at least one α, say α_0, such that $\zeta \in G_{\alpha_0}$. Now, G_{α_0} is open and so there is some $r > 0$ such that $D(\zeta, r) \subseteq G_{\alpha_0}$. But $\zeta \in F_n$, for every n and diam $F_n \to 0$ and so, for all sufficiently large n (in fact, for n with $d/2^{n-1} < r$), we have that $F_n \subseteq D(\zeta, r) \subseteq G_{\alpha_0}$. This contradicts the impossibility of covering any F_n with a finite collection of G_αs—we can do it with just one!

We conclude that there is some finite collection, say $\alpha_1, \ldots, \alpha_m$, such that $K \subseteq G_{\alpha_1} \cup \cdots \cup G_{\alpha_m}$, as required.

For the converse, suppose that any open cover of K possesses a finite subcover. For each $z \in K$, let G_z be the disc $G_z = D(z, 1)$. Evidently, $\{G_z\}_{z \in K}$ is an open cover of K and therefore, by hypothesis, has a finite subcover. That is, there is $z_1, \ldots z_p$ such that

$$K \subseteq D(z_1, 1) \cup \cdots \cup D(z_p, 1).$$

It follows that K is bounded (in fact, $K \subseteq D(0, r)$, where the radius r is given by $r = \max\{ |z_i| : 1 \le i \le p \} + 1$).

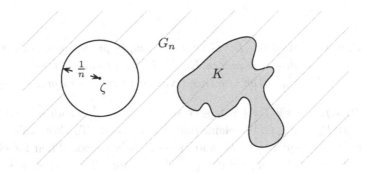

Fig. 3.4 K is closed.

To prove that K is closed, we will show that $\mathbb{C} \setminus K$, the complement of the set K, is open. For this, let $\zeta \in \mathbb{C} \setminus K$ (note that $\mathbb{C} \setminus K$ is not empty because K is bounded). For each $n \in \mathbb{N}$, let $G_n = \{ z : |z - \zeta| > \frac{1}{n} \}$. Then G_n is open, and $G_n \subseteq G_{n+1}$. Furthermore, every $z \ne \zeta$ is contained in some G_n (in fact, z is a member of every G_n for which $n\,|z - \zeta| > 1$) and so certainly $\{G_n\}_{n \in \mathbb{N}}$ is an open cover of K (because $\zeta \notin K$). By hypothesis, there is a finite subcover,

$$K \subseteq G_{n_1} \cup \cdots \cup G_{n_m}$$

for some $n_1, \ldots, n_m \in \mathbb{N}$. Let $k = \max\{n_1, \ldots, n_m\}$. Then $G_{n_i} \subseteq G_k$ for each $i = 1, \ldots, m$ and we have $K \subseteq G_k$. It follows that $D(\zeta, \frac{1}{k}) \subseteq \mathbb{C} \setminus K$ and we conclude that $\mathbb{C} \setminus K$ is open and so the proof is complete. $\qquad \square$

Remark 3.11   We therefore have three equivalent formulations, each one expressing the compactness of a given set K.

- Any sequence in K has a subsequence convergent to some point in the set K.
- K is both closed and bounded.
- Every open cover of K has a finite subcover.

In the more general context of topological spaces, the third version is taken as the definition of compactness. It should be noted that in this more general situation (i.e., in a general topological space) these three statements about the set K are no longer necessarily equivalent. (In fact, boundedness may not even be defined.)

Proposition 3.6   *Let G be a non-empty proper open subset of $\mathbb{C}$. Then there is a nested sequence $K_n \subseteq K_{n+1}$ of compact sets in G such that $G = \bigcup_{n=1}^{\infty} K_n$. (The sequence (K_n) is a compact exhaustion of G.)*

Proof.   Let $F_n = \{\, z : z \in G \text{ and } |z - w| \geq \frac{1}{n} \text{ for all } w \in G^c \,\}$. It is clear that $F_n \subseteq F_{n+1}$. We claim that F_n is closed. To show this, suppose that (z_k) is a sequence in F_n and $z_k \to \zeta$, as $k \to \infty$. Then for each $w \in G^c$, $|z_k - w| \to |\zeta - w|$. But $|z_k - w| \geq \frac{1}{n}$ for all k and so $|\zeta - w| \geq \frac{1}{n}$. In particular, $\zeta \notin G^c$ and so $\zeta \in G$. But then this means that $\zeta \in F_n$ and so F_n is closed.

Let $K_n = F_n \cap \overline{D(0, n)}$. Certainly, $K_n \subseteq G$ and K_n is bounded. K_n is also closed because it is the intersection of closed sets. Hence K_n is compact. It is also clear that $K_n \subseteq K_{n+1}$.

Finally, suppose that $z \in G$. Since G is open, there is $r > 0$ such that $D(z, r) \subseteq G$. In particular, for any $w \in G^c$, $|z - w| \geq r$. Choose $n \in \mathbb{N}$ such that $n > 1/r$ and $n > |z|$. Then $z \in F_n$ and also $z \in \overline{D(0, n)}$, that is, $z \in K_n$. It follows that $G = \bigcup_{n=1}^{\infty} K_n$, as required.   □

By considering the discs $D(0, 1)$ and $D(2i, 1)$ or even $D(0, 1)$ and the closed disc $\overline{D(2i, 1)}$, for example, we see that it is quite possible for two sets A and B to be disjoint but be "almost touching" in the sense that there are points $a \in A$ and $b \in B$ as close together as we wish. This cannot happen for closed sets if one of them is compact, as we now show. (We shall give two proofs, illustrating different aspects of compactness.)

Proposition 3.7   *Suppose that K and F are disjoint subsets of $\mathbb{C}$ with K compact and F closed. Then there is some $r > 0$ such that $|z - w| \geq r$ for every $z \in K$ and $w \in F$. In other words, there is some $r > 0$ such that the distance between any point in K and any point in F is not less than r.*

Proof.   Suppose the contrary, that is, for any $r > 0$ there is some $z \in K$ and $w \in F$ with $|z - w| < r$. In particular, for any $n \in \mathbb{N}$, there is $z_n \in K$ and $w_n \in F$ such that $|z_n - w_n| < \frac{1}{n}$ (taking $r = \frac{1}{n}$). By compactness of K, the sequence (z_n) has a convergent subsequence $(z_{n_k})_{k \in \mathbb{N}}$, say, $z_{n_k} \to \zeta$, as $k \to \infty$, with $\zeta \in K$.

We claim that the (sub)sequence $(w_{n_k})_{k \in \mathbb{N}}$ also converges to ζ. This is to be expected. After all, for large n, the w_ns are close to the z_ns, so the w_{n_k}s get dragged along with the z_{n_k}s towards ζ. To show this, let $\varepsilon > 0$ be given. There is $k_0 \in \mathbb{N}$ such that

$$|\zeta - z_{n_k}| < \tfrac{1}{2}\varepsilon$$

whenever $k > k_0$. Choose $N \in \mathbb{N}$ such that $N > k_0 + \frac{2}{\varepsilon}$. We have

$$|\zeta - w_{n_k}| \leq |\zeta - z_{n_k}| + |z_{n_k} - w_{n_k}|$$
$$< |\zeta - z_{n_k}| + \frac{1}{n_k}$$
$$< \tfrac{1}{2}\varepsilon + \tfrac{1}{2}\varepsilon$$

whenever $k > N$ (since $n_k \geq k$ and $\frac{1}{k} < \frac{1}{2}\varepsilon$, in this case). It follows that $w_{n_k} \to \zeta$, as $k \to \infty$, as claimed. However, each $w_{n_k} \in F$ and, by hypothesis, F is closed. It follows that $\zeta \in F$, that is, ζ is a common point of both K and F. This contradicts the hypothesis that K and F are disjoint. We conclude that there must be $r > 0$ such that $|z - w| \geq r$ for any $z \in K$ and any $w \in F$.

Alternative Proof. By hypothesis, K and F are disjoint. This means that $K \subseteq \mathbb{C} \setminus F$. Since F is a closed set, $G \equiv \mathbb{C} \setminus F$ is open. Hence, for every $z \in K$, there is some $r_z > 0$ such that $D(z, r_z) \subseteq G$. The family $\{ D(z, \frac{1}{2} r_z) : z \in K \}$ of open discs is clearly an open cover of the compact set K (each point of K is the centre of one of the discs). By compactness, this family contains a finite subcover, so it follows that there is a finite set of points $z_1, \ldots, z_m$ in K such that the discs $D(z_1, \frac{1}{2} r_{z_1}), \ldots, D(z_m, \frac{1}{2} r_{z_m})$ cover K.

Set $r = \min\{r_{z_1}, \ldots, r_{z_m}\}$ so that $r > 0$ and consider the collection $\{\, D(z, \frac{1}{2}r) : z \in K \,\}$ of open discs, labelled by the points of K. We claim that for any $z \in K$, the disc $D(z, \frac{1}{2}r)$ belongs to one of the m discs $D(z_j, r_{z_j})$, $j = 1, \ldots, m$. To see this, first note that for given $z \in K$ there is $1 \le j \le m$ such that $z \in D(z_j, \frac{1}{2}r_{z_j})$ (since these m discs cover K). Let w be any point in the disc $D(z, \frac{1}{2}r)$. Then

$$
\begin{aligned}
|w - z_j| &\le |w - z| + |z - z_j| \\
&< \tfrac{1}{2}r + \tfrac{1}{2}r_{z_j} \\
&\le \tfrac{1}{2}r_{z_j} + \tfrac{1}{2}r_{z_j} \\
&= r_{z_j}
\end{aligned}
$$

which means that $D(z, \frac{1}{2}r) \subseteq D(z_j, r_{z_j})$, as claimed.

Each disc $D(z_j, r_{z_j})$ lies in G and so we see that, for every $z \in K$, the disc $D(z, \frac{1}{2}r)$ also lies in G. (This is also true of their union, which contains K; $K \subset \bigcup_{z \in K} D(z, \frac{1}{2}r) \subseteq G$.) Let $w \in F$. Then $w \notin G = \mathbb{C} \setminus F$ and so, in particular, it follows that $w \notin D(z, \frac{1}{2}r)$ and therefore $|z - w| \ge \frac{1}{2}r$, for any $z \in K$. $\qquad \square$

Example 3.12 Let $F_1 = \{\, z = x + iy : x \ge 1 \text{ and } xy \ge 1 \,\}$ and let $F_2 = \{\, z = x - iy : x \ge 1 \text{ and } xy \ge 1 \,\}$. ($F_2$ is the set of complex conjugates of points in F_1, that is, its mirror image in the real-axis. We can also think of F_1 as all those points (x, y) in the plane with $x \ge 1$ and lying above the hyperbola $y = 1/x$.) Evidently, $F_1 \cap F_2 = \varnothing$.

We claim that F_1 and F_2 are closed. To see this, suppose that (z_n) is a sequence in F_1 such that $z_n \to w$. Let $z_n = x_n + iy_n$ and $w = u + iv$. Then we know that $x_n \to u$ and that $y_n \to v$ and so $x_n y_n \to uv$. But $x_n y_n \ge 1$, for all n, and therefore $uv \ge 1$. Furthermore, $x_n \ge 1$ and so $u \ge 1$. It follows that $w = u + iv \in F_1$ and we conclude that F_1 is closed. Similarly, we see that F_2 is closed.

Now, $\zeta_n = n + \frac{1}{n}i \in F_1$ and $\overline{\zeta_n} \in F_2$ and the distance between these two points is $|\zeta_n - \overline{\zeta_n}| = \frac{2}{n}$, which can be made as small as we wish by choosing n sufficiently large. It follows that $\inf\{\, |z - w| : z \in F_1,\ w \in F_2 \,\}$, the distance between the two closed, disjoint sets F_1 and F_2, is zero.

3.8 Polygons and Paths in $\mathbb{C}$

Definition 3.11   The line segment from z_0 to z_1 is the set

$$[z_0, z_1] = \{\, z : z = z_0 + t(z_1 - z_0),\ 0 \le t \le 1 \,\}$$

It consists of all those points in the complex plane lying on the straight line between z_0 and z_1.

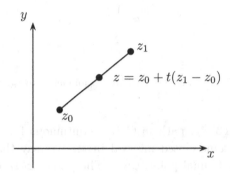

Fig. 3.5   The line segment joining z_0 to z_1.

Remark 3.12   By looking at the real and imaginary parts, we can express this in terms of (familiar) cartesian coordinates. Let $z = x + iy \in [z_0, z_1]$ and write $z_0 = x_0 + iy_0$ and $z_1 = x_1 + iy_1$. z can be written as $z = z_0 + t(z_1 - z_0)$ for some $0 \le t \le 1$, which gives the pair of equations

$$x = x_0 + t(x_1 - x_0)$$
$$y = y_0 + t(y_1 - y_0).$$

Eliminating t shows that (x, y) lies on the straight line

$$y = y_0 + \frac{(y_1 - y_0)}{(x_1 - x_0)}(x - x_0).$$

Moreover, the restriction $0 \le t \le 1$ means that x varies between x_0 (when $t = 0$) and x_1 (when $t = 1$).

Definition 3.12   A polygon in $\mathbb{C}$ is a set of the form

$$[z_0, z_1] \cup [z_1, z_2] \cup \cdots \cup [z_{n-1}, z_n]$$

for some $n \in \mathbb{N}$ and points $z_0, z_1, \ldots, z_n$.

In other words, a polygon is a finite collection of line segments placed end to end. We say that the polygon joins the points z_0 and z_n. Notice that z_n need not be the same as z_0.

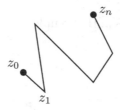

Fig. 3.6   A polygon joining z_0 to z_n.

Definition 3.13   A path in $\mathbb{C}$ is a continuous function $\varphi : [a, b] \to \mathbb{C}$, where $[a, b]$ is some closed interval in $\mathbb{R}$. We say that φ joins its initial point, $\varphi(a)$, to its final point, $\varphi(b)$. The points $\varphi(a)$ and $\varphi(b)$ are its endpoints. The trace (also called the track, or the impression) of the path φ is defined to be the set $\{z : z = \varphi(t), t \in [a, b]\}$ and is denoted by $\operatorname{tr}\varphi$. (Note that, by definition, a map ψ from a subset S of $\mathbb{R}$ to $\mathbb{C}$ is continuous at the point $x_0 \in S$ if for any given $\varepsilon > 0$ there is $\delta > 0$ such that $|x - x_0| < \delta$, with $x \in S$, implies that $|\psi(x) - \psi(x_0)| < \varepsilon$. ψ is said to be continuous on S if it is continuous at each point of S. This is the natural definition of continuity of a complex-valued function of a real variable.)

So a path is a continuous complex-valued function of a real variable, defined on some closed interval in $\mathbb{R}$, and its track is just its range in $\mathbb{C}$.

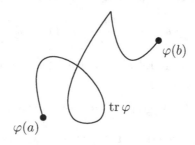

Fig. 3.7   A path joining $\varphi(a)$ to $\varphi(b)$.

Remark 3.13 A polygon can be thought of as a path, as follows. For given points (vertices) $z_0, \ldots, z_n \in \mathbb{C}$, define $\varphi : [0, n] \to \mathbb{C}$ by $\varphi(n) = z_n$ and

$$\varphi(t) = z_k + (t - k)(z_{k+1} - z_k) \quad \text{for } k \leq t < k + 1$$

for $k = 0, 1, 2, \ldots, n - 1$. Then as t increases from k to $k + 1$, $\varphi(t)$ moves along the line segment $[z_k, z_{k+1}]$ from z_k towards z_{k+1}. Evidently, the trace of φ is given by $\operatorname{tr} \varphi = [z_0, z_1] \cup [z_1, z_2] \cup \cdots \cup [z_{n-1}, z_n]$.

Remark 3.14 Let $\varphi : [a, b] \to \mathbb{C}$ be a given path. Since φ is continuous, it follows that $\operatorname{tr} \varphi$ is compact in $\mathbb{C}$. To see this, let (z_n) be any sequence in $\operatorname{tr} \varphi$. Then $z_n = \varphi(t_n)$ for some $t_n \in [a, b]$. But (t_n) has a convergent subsequence $t_{n_k} \to \tau$, say, as $k \to \infty$, with $\tau \in [a, b]$. The continuity of φ implies that $\varphi(t_{n_k}) \to \varphi(\tau)$. Thus, $z_{n_k} = \varphi(t_{n_k}) \to \varphi(\tau) \in \operatorname{tr} \varphi$, as $k \to \infty$ and so $\operatorname{tr} \varphi$ is compact. We note, in particular, that any line segment $[z_0, z_1]$ is compact in $\mathbb{C}$.

Remark 3.15 Suppose that $\varphi : [a, b] \to \mathbb{C}$ and $\psi : [c, d] \to \mathbb{C}$ are paths with $\varphi(b) = \psi(c)$, i.e., ψ starts where φ ends. We can join these together into a single path as follows. Define $\gamma : [a, b + d - c] \to \mathbb{C}$, by the formula

$$\gamma(t) = \begin{cases} \varphi(t), & a \leq t < b \\ \psi(c + t - b), & b \leq t \leq b + d - c. \end{cases}$$

We see that γ is a path starting from $\varphi(a)$ and ending at $\psi(d)$. If $c = b$, then γ is equal to φ on the subinterval $[a, b]$ of $[a, d]$ and equal to ψ on $[b, d]$. In any event, we shall write $\gamma = \varphi + \psi$.

3.9 Connectedness

Definition 3.14 A subset $S \subseteq \mathbb{C}$ is said to be pathwise connected if for any pair of points z_0, z_1 in S there is some path, φ, say, lying entirely in S (i.e., such that $\operatorname{tr} \varphi \subseteq S$) and joining z_0 to z_1

Definition 3.15 A set A in $\mathbb{C}$ is said to be polygonally connected if and only if for any two points $z', z'' \in A$ there is some polygon lying entirely within A which joins z' and z''.

Definition 3.16 A set A in $\mathbb{C}$ is said to be stepwise connected if and only if for any two points $z', z'' \in A$ there is some polygon with line segments parallel to either the real or the imaginary axes which lies entirely within A and which joins z' and z''.

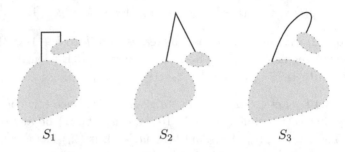

$$S_1 \qquad\qquad S_2 \qquad\qquad S_3$$

Fig. 3.8 Stepwise, polygonally and pathwise connected sets.

Figure 3.8 illustrates these three notions of connectedness. (Each set comprises two "blobs" (without their boundary) attached by a line.) Notice that S_3 is not polygonally connected, and S_2 is not stepwise connected.

Remark 3.16 The notion of connectedness is supposed to convey the idea of a set being "all in one piece". Clearly, the circle $\{\, z : |z| = 1 \,\}$ is "all one piece" and so should qualify as being connected. However, evidently it is *not* polygonally connected—because it contains no line segments at all (it is everywhere curved!). One could therefore wonder at the usefulness of the notion of polygonal connectedness. It turns out that this is very convenient in the consideration of open sets—which is of special importance in complex analysis. (The precise point here is that pathwise connectedness is equivalent to polygonal connectedness for open sets in $\mathbb{C}$, as we will soon show.)

There is a further notion of connectedness as follows.

Definition 3.17 A subset S of $\mathbb{C}$ is said to be disconnected if there are disjoint open sets A and B (so $A \cap B = \varnothing$) such that $A \cap S \neq \varnothing$, $B \cap S \neq \varnothing$ and $S \subseteq A \cup B$. A set is said to be connected if it is not disconnected.

For open sets, the situation is a little less complicated, as the following proposition shows.

Proposition 3.8 *Let G be an open subset of $\mathbb{C}$. Then G is disconnected if and only if there are non-empty disjoint open sets G_1 and G_2, say, such that $G = G_1 \cup G_2$.*

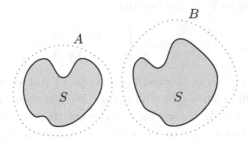

Fig. 3.9   The set S (shaded) is disconnected (the open sets A and B are dotted).

Proof.   If G_1 and G_2 exist as stated, then by definition, G is disconnected.

Conversely, suppose that G is disconnected; $G \subseteq A \cup B$, with A, B disjoint, open sets such that $A \cap G \neq \varnothing$ and $B \cap G \neq \varnothing$. Setting $G_1 = A \cap G$ and $G_2 = B \cap G$ gives the open sets required. $\qquad\square$

The next result is geometrically obvious, but nevertheless, requires proof. It rests on the completeness property of the real number system (which plays a central role in the Intermediate Value Theorem).

Theorem 3.6   *Suppose that A is a proper non-empty subset of $\mathbb{C}$. Then any path joining a point of A to a point not in A contains a point of the boundary of A. In other words, for a path to "escape" from a set, it must cross its boundary.*

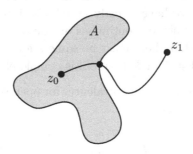

Fig. 3.10   The path must cross the boundary of A.

Proof.   Suppose that $\varphi : [a, b] \to \mathbb{C}$ is a path joining $z_0 \in A$ to $z_1 \notin A$.

Let $g : [a, b] \to \mathbb{R}$ be the function

$$g(t) = \begin{cases} 0, & \varphi(t) \in A \\ 1, & \varphi(t) \notin A. \end{cases}$$

The function g maps $[a, b]$ onto the points $\{0, 1\}$ and so, by the Intermediate Value Theorem, g cannot be continuous. We will show that g is continuous at every point s in $[a, b]$ for which $\varphi(s) \notin \partial A$. If φ did not meet ∂A then g would be continuous on the whole of $[a, b]$, which we know to be false.

To proceed, then, suppose that $s \in [a, b]$ and that $\varphi(s) \notin \partial A$. Suppose first that $\varphi(s) \in A$, so that $g(s) = 0$. Since $\varphi(s) \notin \partial A$, there is $r > 0$ such that $D(\varphi(s), r) \subseteq A$. By continuity of φ, there is $\delta > 0$ such that

$$|t - s| < \delta, \quad t \in [a, b] \implies \varphi(t) \in D(\varphi(s), r) \subseteq A,$$

that is, $g(t) = 0 = g(s)$. Hence, for any given $\varepsilon > 0$,

$$|g(t) - g(s)| = |0 - 0| = 0 < \varepsilon$$

whenever $|t - s| < \delta$, $t \in [a, b]$. In other words, g is continuous at s.

Now suppose that $\varphi(s) \notin A$. Then $g(s) = 1$. Arguing just as before, we conclude that there is $r' > 0$ such that $D(\varphi(s), r') \subseteq \mathbb{C} \setminus A$ and that there is some $\delta' > 0$ such that

$$\varphi(t) \in D(\varphi(s), r') \subseteq \mathbb{C} \setminus A$$

whenever $|t - s| < \delta'$, $t \in [a, b]$. In other words, $g(t) = 1$ whenever we have $|t - s| < \delta'$, $t \in [a, b]$, which again shows that g is continuous at s.

We have shown that g is continuous at every point s in $[a, b]$ such that $\varphi(s) \notin \partial A$. As we have already noted, the Intermediate Value Theorem implies that g cannot be continuous on the whole of the interval $[a, b]$ and so we conclude that there must be some s for which $\varphi(s) \in \partial A$. In other words, $\operatorname{tr} \varphi \cap \partial A \neq \varnothing$ and the proof is complete. $\qquad\square$

We can now prove the equivalence, for open sets, of the various notions of connectedness.

Theorem 3.7   *Let G be an open set. The following are equivalent.*

  (i) *G is stepwise connected.*
 (ii) *G is polygonally connected.*
(iii) *G is pathwise connected.*
(iv) *G is connected.*

Proof. If $G = \varnothing$ there is nothing to prove, since $\varnothing$ is "connected" in each of the four senses anyway.

Suppose that $G \neq \varnothing$. Clearly, (i) $\implies$ (ii) $\implies$ (iii). We first show that (iii) $\implies$ (iv). Suppose then, that G is a non-empty pathwise connected open set in $\mathbb{C}$.

By way of contradiction, suppose that G is not connected, i.e., that G is disconnected. Then by proposition 3.8 there are disjoint, non-empty open sets A and B such that $G = A \cup B$. Let $z_0 \in A$ and suppose that $z_1 \in B$. By hypothesis, there is a path φ joining z_0 to z_1 in G. Since $z_1 \notin A$, the path φ passes through the boundary of A. That is, there is some $z^* \in \operatorname{tr} \varphi$ with $z^* \in \partial A$. Since A is open, $z^* \notin A$ and therefore $z^* \in B$. But then $D(z^*, r) \subseteq B$ for some $r > 0$. In particular, $D(z^*, r) \cap A = \varnothing$, which contradicts $z^* \in \partial A$. We deduce that G is connected, as required.

Notice that by replacing "path" by "step-path" or by "polygonal path", respectively, it follows that (iv) is a direct consequence of any one of the properties (i), (ii) or (iii).

To show that (iv) $\implies$ (i), and hence complete the proof, suppose that G is open and connected. Let z_0 and z_1 belong to G and set

$$A = \{ \zeta \in G : \zeta \text{ is stepwise connected to } z_0 \text{ in } G \}$$
$$B = G \setminus A.$$

We shall show that A is open and non-empty. Since G is open, there is $r > 0$ such that $D(z_0, r) \subseteq G$. Evidently, $D(z_0, r) \subseteq A$, so $A \neq \varnothing$.

Let $w \in A$. Then $w \in G$ and so $D(w, \rho) \subseteq G$ for some $\rho > 0$. Now, any point in $D(w, \rho)$ is stepwise connected to z_0, via w, and so $D(w, \rho) \subseteq A$ and it follows that A is open.

Next we show that B is also open (in fact, we will see that B is empty). If $z' \in B$, then $z' \in G$ and so $D(z', r') \subseteq G$, for some $r' > 0$. Any point w, say, in $D(z', r')$ is stepwise connected to z'. If w were stepwise connected to z_0, then z' would be as well (via w). We conclude that no point of $D(z', r')$ can be stepwise connected to z_0, i.e., $D(z', r') \subseteq B$, showing that B is open. But $G = A \cup B$ and G is connected, by hypothesis. It follows that $B = \varnothing$. Hence $G = A$ and the proof is complete. $\square$

What's going on? The equivalence of these various notions of a set being all in one piece is only claimed to hold for *open* sets in $\mathbb{C}$. They are not the same in general. However, our main concern will be with open sets, so we can appeal to whichever version we find appropriate at any particular time, to suit our own convenience.

We can now easily answer the question "which sets are both open and closed"?

Theorem 3.8   *The only subsets of $\mathbb{C}$ which are simultaneously open and closed are $\varnothing$ and $\mathbb{C}$ itself.*

Proof.   By way of contradiction, suppose that A is a non-empty proper subset of $\mathbb{C}$ and that A is both open and closed. It follows, in particular, that $\mathbb{C} \setminus A$, the complement of A, is open. But then we can write $\mathbb{C}$ as the disjoint union of two open sets A and $B = \mathbb{C} \setminus A$, $\mathbb{C} = A \cup B$. Since A is a proper subset of $\mathbb{C}$ it follows that B is non-empty. This means that $\mathbb{C}$ is disconnected.

However, $\mathbb{C}$ is open and evidently polygonally connected (indeed, $[z_0, z_1]$ joins the points z_0 and z_1). But this means that $\mathbb{C}$ is also connected, by theorem 3.7. This gives a contradiction and we deduce that no such non-empty proper subset A exists.   □

3.10   Domains

Definition 3.18   An open connected set is called a domain (also sometimes known as a region).

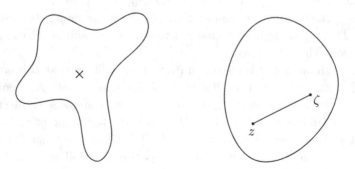

Fig. 3.11   Examples of a star-domain and a convex domain.

Definition 3.19   A domain D is (star-like or) a star-domain if there is some $z_0 \in D$ such that for each $z \in D$ the line segment $[z_0, z]$ lies in D. Any such point z_0 is called a star-centre.

A domain D is convex if for any pair of points $z, \zeta \in D$, the line segment $[z, \zeta]$ lies in D.

Evidently, if D is convex, then it is star-like and each of its points is a star-centre. The converse is false, in general.

Example 3.13 The set $D = \{\, z : z + |z| \neq 0 \,\}$ is star-like, but not convex. In fact, D is the whole complex plane with the negative real-axis (including the point 0) removed. (The complement of D is the set $\{\, z : z = -|z| \,\}$, which is the set of complex numbers which are real and negative (or 0).) D is not convex because the line segment joining, say, the point $(-1 - i)$ to the point $(-1 + i)$ crosses the negative real-axis at $z = -1$ which does not belong to D. On the other hand, we see that, for example, $z_0 = 1$ is a star-centre. (For any $z \in D$, the line segment $[1, z]$ lies entirely in D.)

Example 3.14 Let L_1 and L_2 be the semi-infinite line segments given by $L_1 = \{\, z : z = r, \quad r \geq 1 \,\}$ and $L_2 = \{\, z : z = ir, \quad r \geq 1 \,\}$ and let $D = \mathbb{C} \setminus (L_1 \cup L_2)$. Evidently D is a star-domain with star-centre $z_0 = 0$. Moreover, the point $z_0 = 0$ is the only star-centre for D.

Example 3.15 For $0 < r < R$, the ring (annulus) $A = \{\, z : r < |z| < R \,\}$ is pathwise connected. To see this, let z_1 and z_2 be any pair of points in the annulus A. Let $\rho_1 = |z_1|$ and $\rho_2 = |z_2|$. Then z_1 can be connected to $w_1 = \rho_1$ (on the positive real axis) in A by the path given by $\varphi_1(t) = \rho_1\big(\cos((1 - t)\operatorname{Arg} z_1) + i\sin((1 - t)\operatorname{Arg} z_1)\big), 0 \leq t \leq 1$.

The point w_1 can be joined to the point $w_2 = \rho_2$ (also on the positive real axis) in A by the path $\varphi_2(t) = w_1 + t(w_2 - w_1), 0 \leq t \leq 1$.

Finally, the point w_2 can be joined to z_2 in A by means of the path $\varphi_3(t) = \rho_2\big(\cos(t\operatorname{Arg} z_2) + i\sin(t\operatorname{Arg} z_2)\big), 0 \leq t \leq 1$.

Combining these three paths, we get a path which lies in the annulus A and joins z_1 to z_2, as required.

Example 3.16 Suppose D_1 and D_2 are domains and that $D_1 \cap D_2 \neq \varnothing$. Then $D_1 \cup D_2$ is also a domain. Clearly $D_1 \cup D_2$ is open. To see that $D_1 \cup D_2$ is pathwise connected, let z_1 and z_2 be any two points in $D_1 \cup D_2$ and let w be any point in $D_1 \cap D_2$. By hypothesis, there is a path $\varphi_1 : [0, 1] \to \mathbb{C}$ joining z_1 to w in D_1 and a path $\varphi_2 : [0, 1] \to \mathbb{C}$ joining w to z_2 in D_2. For $0 \leq t \leq 1$, set

$$\varphi(t) = \begin{cases} \varphi_1(2t), & 0 \leq t \leq \tfrac{1}{2} \\ \varphi_2(2t - 1), & \tfrac{1}{2} < t \leq 1. \end{cases}$$

Then φ is a path joining z_1 to z_2 (via w) in $D_1 \cup D_2$ which shows that the set $D_1 \cup D_2$ is connected.

Notice that, in general, the union $D_1 \cup D_2$ need not be star-like even if both D_1 and D_2 are. For example, let $D_1 = \mathbb{C} \setminus L_1$ and $D_2 = \mathbb{C} \setminus L_2$ where $L_1 = \{ z : z = x + iy, \ x \geq 0, \ y = 0 \}$ is the non-negative real-axis and $L_2 = \{ z : z = x + iy, \ x \leq 0, \ y = 0 \}$ is the non-positive real-axis. Then D_1 and D_2 are both star-like but $D_1 \cup D_2 = \mathbb{C} \setminus \{ 0 \}$, the punctured plane, which is not star-like.

It is natural to ask whether $D_1 \cap D_2$ is a domain if both D_1 and D_2 are. This need not be the case. For example, let D_1 be the ring (annulus) given by $D_1 = \{ z : 10 < |z| < 11 \}$ and let D_2 be the vertical strip given by $D_2 = \{ z : -1 < \operatorname{Re} z < 1 \}$. Evidently both D_1 and D_2 are domains but $D_1 \cap D_2$ consists of two disconnected parts.

Chapter 4

Analytic Functions

4.1 Complex-Valued Functions

A complex-valued function of a complex variable is a mapping f, say, from (a subset of) $\mathbb{C}$ into $\mathbb{C}$. The mapping f is real-valued if its range lies in $\mathbb{R}$ (considered as a subset of $\mathbb{C}$). Note that f need not be defined on the whole of $\mathbb{C}$; for example, we may wish to consider the function $z \mapsto f(z) = \frac{1}{z}$ which is undefined at $z = 0$. We will be concerned almost exclusively with functions defined on domains in $\mathbb{C}$, but as we have already seen, we will also need to consider complex-valued functions of a real variable, such as paths.

4.2 Continuous Functions

The definitions of continuity and of differentiability are straightforward extensions of the corresponding real versions.

Definition 4.1 A function $f : A \to \mathbb{C}$ is continuous at $z_0 \in A$ if and only if for any given $\varepsilon > 0$ there is $\delta > 0$ such that $|z - z_0| < \delta$ and $z \in A$ imply that $|f(z) - f(z_0)| < \varepsilon$.

In words, f is continuous at z_0 if and only if $f(z)$ is as close to $f(z_0)$ as we wish, provided z is sufficiently close to z_0.

Remark 4.1 The set A is assumed to be a given subset of $\mathbb{C}$, but by assuming it to be a subset of $\mathbb{R}$, we get the definition of continuity of a complex-valued function of a real variable. If, in addition, f happens to be real-valued, then we recover precisely the definition of continuity of a real function of a real variable. (This is because the modulus on $\mathbb{C}$ extends that on $\mathbb{R}$.)

The following result is sometimes useful. (One might prefer to think in terms of sequences rather than ε-δs.)

Proposition 4.1   *The function $f : A \to \mathbb{C}$ is continuous at $z_0 \in A$ if and only if $f(\zeta_n) \to f(z_0)$ for every sequence (ζ_n) in A which converges to z_0.*

Proof.   Suppose that f is continuous at $z_0 \in A$ and that (ζ_n) is any sequence in A which converges to z_0. Let $\varepsilon > 0$ be given. Then there is $\delta > 0$ such that $|f(z) - f(z_0)| < \varepsilon$ whenever $|z - z_0| < \delta$, $z \in A$. However, there is $N \in \mathbb{N}$ such that $|\zeta_n - z_0| < \delta$ whenever $n > N$. It follows that $|f(\zeta_n) - f(z_0)| < \varepsilon$ whenever $n > N$, i.e., $(f(\zeta_n))$ converges to $f(z_0)$, as $n \to \infty$.

Conversely, suppose that $f(\zeta_n) \to f(z_0)$ whenever $\zeta_n \to z_0$, with $\zeta_n \in A$. Suppose that f is not continuous at z_0. This means that there is some $\varepsilon_0 > 0$ such that no matter what $\delta > 0$ is, there is some $\zeta \in A$ with $|\zeta - z_0| < \delta$ such that $|f(\zeta) - f(z_0)| \geq \varepsilon_0$.

In particular, for each $n \in \mathbb{N}$, there is some element of A, ζ_n, say, such that $|\zeta_n - z_0| < 1/n$ but $|f(\zeta_n) - f(z_0)| \geq \varepsilon_0$. Evidently, we have a sequence (ζ_n) in A such that $\zeta_n \to z_0$, as $n \to \infty$, but such that $(f(\zeta_n))$ does not converge to $f(z_0)$. This contradicts our hypothesis, and so we conclude that f is continuous at z_0, and the proof is complete. $\square$

Some basic facts about continuous functions can be readily established.

Proposition 4.2

(i) *Suppose that $f : A \to \mathbb{C}$ is continuous at $z_0 \in A$. Then so are the functions $\operatorname{Re} f$, $\operatorname{Im} f$, $\overline{f}$ and $|f|$.*

(ii) *Suppose that $f : A \to \mathbb{C}$ and $g : A \to \mathbb{C}$ are continuous at $z_0 \in A$. Then so is their sum $f + g$ and their product fg.*

(iii) *Suppose that $f : A \to \mathbb{C}$ and $h : B \to \mathbb{C}$ where $f(z) \in B$ for all $z \in A$. If f is continuous at $z_0 \in A$ and h is continuous at $f(z_0)$, then the composition $h \circ f : A \to \mathbb{C}$ is continuous at z_0.*

Proof.   We shall use proposition 4.1. Suppose (ζ_n) is any sequence in A which converges to z_0. Then, by proposition 4.1, $(f(\zeta_n))$ converges to $f(z_0)$ and $(g(\zeta_n))$ converges to $g(z_0)$. Therefore $\operatorname{Re} f(\zeta_n) \to \operatorname{Re} f(z_0)$, $\operatorname{Im} f(\zeta_n) \to \operatorname{Im} f(z_0)$, $\overline{f(\zeta_n)} \to \overline{f(z_0)}$, $|f(\zeta_n)| \to |f(z_0)|$, $(f + g)(\zeta_n) \to (f + g)(z_0)$ and $(fg)(\zeta_n) \to (fg)(z_0)$. Parts (i) and (ii) now follow, once again, by proposition 4.1.

Furthermore, by proposition 4.1, $f(\zeta_n) \to f(z_0)$ implies that $h(f(\zeta_n)) \to h(f(z_0))$, as $n \to \infty$. Again, by proposition 4.1, we see that the composition $h \circ f$ is continuous at z_0, which proves (iii). $\qquad\square$

Of course, these results could have been proved directly from the ε-δ definition of continuity. We will need the following result later.

Proposition 4.3 *Suppose K is a compact set in $\mathbb{C}$ and $f : K \to \mathbb{C}$ is continuous. Then the image $f(K)$ is compact. In particular, if γ is any path and if f is continuous on $\operatorname{tr}\gamma$, then there is $M > 0$ such that $|f(w)| \leq M$ for all $w \in \operatorname{tr}\gamma$.*

Proof. Let (ζ_n) be any sequence in $f(K)$. Then for each n, there is $z_n \in K$ with $f(z_n) = \zeta_n$. Since K is compact, there is a convergent subsequence $z_{n_k} \to z \in K$ as $k \to \infty$. By continuity of f, it follows that $f(z_{n_k}) \to f(z)$ as $k \to \infty$, that is, $\zeta_{n_k} \to \zeta \equiv f(z) \in f(K)$ and so $f(K)$ is compact.

To prove last part, we note that the trace $\operatorname{tr}\gamma$ of any path γ is compact (see remark 3.14). Hence the set $f(\operatorname{tr}\gamma)$ is compact and so, in particular, it is bounded and the proof is complete. $\qquad\square$

4.3 Complex Differentiable Functions

Definition 4.2 Let D be an open set and suppose $f : D \to \mathbb{C}$. f is differentiable at $z_0 \in D$ if and only if there is $\zeta_0 \in \mathbb{C}$ such that for any given $\varepsilon > 0$ there is some $\delta > 0$ such that

$$\left| \frac{f(z) - f(z_0)}{z - z_0} - \zeta_0 \right| < \varepsilon \tag{4.1}$$

for all $z \in D$ satisfying $0 < |z - z_0| < \delta$. The complex number ζ_0 is the derivative of f at the point $z_0 \in D$ and is written $f'(z_0) = \zeta_0$.

In other words, f is differentiable at z_0 if and only if there is some $\zeta_0 \in \mathbb{C}$ such that $\big(f(z) - f(z_0)\big)/(z - z_0) \to \zeta_0$ as $z \to z_0$ with $z \neq z_0$.

Note that $z = z_0$ is not allowed because of the term $z - z_0$ which appears in the denominator. We also observe that since D is open, there is some $r > 0$ such that the disc $D(z_0, r)$ is contained in D and so the left hand side in Eq. (4.1) is well-defined for all $0 < |z - z_0| < r$.

Remark 4.2 This definition of differentiability of a complex function of a complex variable is the straightforward extension of that for a real function of a real variable. On the basis of this, one might expect that the theory of complex differentiable functions is much the same as that of real differentiable functions. This is far from true. The theory of complex differentiable functions is a much "tighter" theory. A hint of this can be glimpsed directly from the definition. In order for the function f to be differentiable at z_0, the limit $(f(z) - f(z_0))/(z - z_0)$ must exist, as $z \to z_0$, no matter *how* z approaches the complex number z_0. The point is that z can approach z_0 in many ways (for example, from various directions, in a spiral, or quite haphazardly). Nonetheless, the quotients $(f(z) - f(z_0))/(z - z_0)$ are required to always have a limit, and this limit is always to be the same, namely ζ_0. In the case of a real variable, the real number x can only approach the real number x_0 from the left or from the right (or a combination of both). So, in a sense, it seems that, in the real case, it might be easier for differentiability to be realized. Put another way, we might expect that in order for a complex function of a complex variable to be differentiable, it ought to be better-behaved than its real counterpart.

This is, indeed, the case, and we will see as the theory unfolds, that complex differentiability has far reaching consequences.

The next result is no surprise.

Proposition 4.4 *Suppose f is differentiable at the point $z_0 \in \mathbb{C}$. Then f is continuous at z_0.*

Proof. Let $\varepsilon > 0$ be given. Then we know that there is some $\delta > 0$ such that

$$\left| \frac{f(z) - f(z_0)}{z - z_0} - f'(z_0) \right| < \varepsilon$$

whenever $0 < |z - z_0| < \delta$. It follows that

$$|f(z) - f(z_0) - f'(z_0)(z - z_0)| < \varepsilon |z - z_0|$$

whenever $0 < |z - z_0| < \delta$. Hence, for any such z,

$$\begin{aligned}
|f(z) - f(z_0)| &= |f(z) - f(z_0) - f'(z_0)(z - z_0) + f'(z_0)(z - z_0)| \\
&\le |f(z) - f(z_0) - f'(z_0)(z - z_0)| + |f'(z_0)(z - z_0)| \\
&< \varepsilon |z - z_0| + |f'(z_0)| \, |z - z_0|.
\end{aligned}$$

Let $\delta' = \min\{\delta, 1/2, \varepsilon/(2(|f'(z_0)| + 1))\}$. Then $|f(z) - f(z_0)| < \varepsilon$ whenever $|z - z_0| < \delta'$ and the proof is complete. $\square$

The usual basic properties of differentiation hold, as shown in the next proposition.

Proposition 4.5

(i) *Suppose $f : A \to \mathbb{C}$ and $g : A \to \mathbb{C}$ are differentiable at $z_0 \in A$. Then so are $f + g$ and fg with derivatives*

$$(f + g)'(z_0) = f'(z_0) + g'(z_0)$$

and

$$(fg)'(z_0) = f'(z_0)g(z_0) + f(z_0)g'(z_0),$$

respectively. If $f(z) \neq 0$, then the quotient $1/f$ is differentiable at z_0 and $(1/f)'(z_0) = -f'(z_0)/f(z_0)^2$ (Quotient Rule).

(ii) *(Chain Rule) If $f : A \to \mathbb{C}$ is differentiable at $z_0 \in A$, $\operatorname{ran} f \subseteq B$ and $g : B \to \mathbb{C}$ is differentiable at $f(z_0)$, then the composition $g \circ f$ is differentiable at z_0 with derivative*

$$(g \circ f)'(z_0) = g'(f(z_0))f'(z_0).$$

Proof.   For any $z \neq z_0$, we have

$$\left| \frac{(f + g)(z) - (f + g)(z_0)}{z - z_0} - (f'(z_0) + g'(z_0)) \right|$$

$$= \left| \frac{f(z) - f(z_0)}{z - z_0} - f'(z_0) + \frac{g(z) - g(z_0)}{z - z_0} - g'(z_0) \right|$$

$$\leq \left| \frac{f(z) - f(z_0)}{z - z_0} - f'(z_0) \right| + \left| \frac{g(z) - g(z_0)}{z - z_0} - g'(z_0) \right|.$$

The right hand side is arbitrarily small provided $|z - z_0|$ is sufficiently small (and not zero) and so we deduce that $f + g$ is differentiable at z_0 with derivative $(f + g)'(z_0) = f'(z_0) + g'(z_0)$.

Next, consider

$$\frac{f(z)g(z) - f(z_0)g(z_0)}{z - z_0} = \frac{(f(z) - f(z_0))g(z)}{z - z_0} + \frac{f(z_0)(g(z) - g(z_0))}{z - z_0}.$$

Using the continuity of g at z_0, we see that the first term on the right hand side approaches $f'(z_0)g(z_0)$ whilst the second term approaches $f(z_0)g'(z_0)$, as $z \to z_0$, with $z \neq z_0$, as required.

Assuming $f(z) \neq 0$, we see that for $z \neq z_0$,

$$\frac{(1/f)(z) - (1/f)(z_0)}{z - z_0} = \frac{f(z_0) - f(z)}{(z - z_0)f(z_0)f(z)} \rightarrow -\frac{f'(z_0)}{f(z_0)^2}$$

as $z \rightarrow z_0$ (since f is continuous at z_0). This proves the Quotient Rule.
To prove the Chain Rule, let $\Phi : B \rightarrow \mathbb{C}$ be given by

$$\Phi(w) = \begin{cases} \dfrac{g(w) - g(f(z_0))}{w - f(z_0)}, & \text{for } w \neq f(z_0) \\ g'(f(z_0)), & \text{for } w = f(z_0). \end{cases}$$

By hypothesis (namely, that g is differentiable at $f(z_0)$), it follows that Φ is continuous at $w = f(z_0)$. Now, as $z \rightarrow z_0$, $f(z) \rightarrow f(z_0)$ and therefore $\Phi(f(z)) \rightarrow \Phi(f(z_0)) = g'(f(z_0))$. For $z \in A$ with $f(z) \neq f(z_0)$, we can rewrite the definition of Φ to get

$$g(f(z)) - g(f(z_0)) = \Phi(f(z)) \left(f(z) - f(z_0) \right).$$

Clearly this equality also holds for z with $f(z) = f(z_0)$, since in this case both sides vanish. Let $z \neq z_0$ and divide both sides by $z - z_0$ to get

$$\frac{g(f(z)) - g(f(z_0))}{z - z_0} = \Phi(f(z)) \frac{\left(f(z) - f(z_0) \right)}{z - z_0}.$$

Letting $z \rightarrow z_0$, the right hand side converges to $g'(f(z_0))f'(z_0)$ and the result follows. □

Examples 4.1

(1) Let $f(z) = \zeta$, for all $z \in \mathbb{C}$, i.e., f is constant. Clearly f is differentiable with derivative $f'(z) = 0$, $z \in \mathbb{C}$.

(2) Let $f(z) = z$, for $z \in \mathbb{C}$. Let $z_0 \in \mathbb{C}$. Then, for any $z \neq z_0$,

$$\frac{f(z) - f(z_0)}{z - z_0} = \frac{z - z_0}{z - z_0} = 1$$

and so

$$\left| \frac{f(z) - f(z_0)}{z - z_0} - 1 \right| = 0$$

and we conclude that f is differentiable at every z_0, with $f'(z_0) = 1$.

(3) For any $n \in \mathbb{N}$, the function $f(z) = z^n$ is differentiable at every $z \in \mathbb{C}$, with derivative $f'(z) = nz^{n-1}$. This can be proved directly using the binomial theorem, or else proved by induction, using the product rule.

(4) The function $f(z) = \dfrac{1}{z}$ is differentiable at every $z \neq 0$, with derivative $f'(z_0) = -\dfrac{1}{z_0^2}$. Indeed, for $z \neq 0$, $z_0 \neq 0$ and $z \neq z_0$,

$$\frac{1/z - 1/z_0}{z - z_0} = \frac{z_0 - z}{z \, z_0 (z - z_0)} = \frac{-1}{z \, z_0} \to -\frac{1}{z_0^2}$$

as $z \to z_0$.

(5) For any $n \in \mathbb{N}$, the function $f(z) = z^{-n}$ is differentiable at any $z \neq 0$ with derivative $f'(z) = -n/z^{n+1}$. This may be proved by induction.

(6) For $z \in \mathbb{C}$, set $f(z) = \bar{z}$ and let $z_0 \in \mathbb{C}$ be given. Then for any complex number $z = x + iy \neq z_0 = x_0 + iy_0$, we have

$$\frac{f(z) - f(z_0)}{z - z_0} = \frac{\overline{z - z_0}}{z - z_0} = \frac{a - ib}{a + ib} \qquad (*)$$

where we have put $z - z_0 = a + ib$ so that $a = x - x_0$ and $b = y - y_0$. We are interested in the behaviour of the quotient $(*)$ when $|z - z_0|$ becomes small. Let z be such that $\operatorname{Im} z = \operatorname{Im} z_0$. Then $b = 0$ and we see that the quotient $(*)$ is equal to 1 (as long as $z \neq z_0$, i.e., $a \neq 0$). On the other hand, if z is such that $\operatorname{Re} z = \operatorname{Re} z_0$, then $a = 0$ and the quotient $(*)$ is equal to -1 (as long as $z \neq z_0$, i.e., $b \neq 0$). In other words, whenever z lies on the line through z_0 parallel to the real axis, $(*)$ assumes the value 1, whereas whenever z lies on the line through z_0 parallel to the imaginary axis, the value of the quotient $(*)$ is -1. We conclude that the quotient $(*)$ does *not* have a limit as $z \to z_0$. (We have already exhibited two "limits", namely, ± 1.)

(7) Let $f(z) = |z|^2$, for $z \in \mathbb{C}$, and let $z_0 = 0$. Then

$$\frac{f(z) - f(z_0)}{z - z_0} = \frac{|z|^2 - 0}{z - 0} = \frac{z \bar{z}}{z} = \bar{z} \to 0$$

as $z \to z_0 = 0$. Hence f is differentiable at $z = 0$ with $f'(0) = 0$. This function f is not differentiable at any other value of z. Indeed, suppose the contrary, namely, that $f(z) = |z|^2$ is differentiable at some $z_0 \neq 0$. Then the product fg would also be differentiable at z_0, where g is the function $g(z) = 1/z$. But $f(z)g(z) = \bar{z}$, which we have seen is nowhere differentiable. This contradiction establishes our claim that f is not differentiable at any $z_0 \neq 0$.

4.4   Cauchy-Riemann Equations

We can relate complex-valued functions of a complex variable to real-valued functions of two real variables by looking at the real and imaginary parts:

$$z = x + iy \quad \longleftrightarrow \quad (x, y) \in \mathbb{R}^2$$
$$f(x + iy) = \operatorname{Re} f(x + iy) + i \operatorname{Im} f(x + iy)$$
$$\equiv u(x, y) + iv(x, y).$$

That is, given $f(z)$, we define $u, v : \mathbb{R}^2 \to \mathbb{R}$ by $u(x, y) = \operatorname{Re} f(x + iy)$ and $v(x, y) = \operatorname{Im} f(x + iy)$.

Conversely, given two functions $u(x, y)$ and $v(x, y)$, we can construct the function $f(z) = u(x, y) + iv(x, y)$, where $z = x + iy$.

The complex differentiability of f implies that u and v, the real and imaginary parts of f, are not completely independent of each other, as we shall now discuss.

Suppose that f is differentiable at $z_0 = x_0 + iy_0$. Then we know that the quotient $(f(z) - f(z_0))/(z - z_0) \to f'(z_0)$ as $z \to z_0$.

We consider two cases, the first where $z = z_0 + s$, with $s \in \mathbb{R}$. To say that $z \to z_0$ is to say that $s \to 0$. Then

$$\frac{f(z) - f(z_0)}{z - z_0} = \frac{u(x_0 + s, y_0) - u(x_0, y_0) + i(v(x_0 + s, y_0) - v(x_0, y_0))}{s}$$
$$\to f'(z_0),$$

as $s \to 0$, with $s \neq 0$. This means that the real and imaginary parts of the left hand side separately have limits, which is to say that the partial derivatives u_x and v_x exist at (x_0, y_0) and, moreover,

$$u_x(x_0, y_0) + iv_x(x_0, y_0) = f'(z_0).$$

Next, consider $z = z_0 + it$ with $t \in \mathbb{R}$. Once again, we know that

$$\frac{f(z) - f(z_0)}{z - z_0} = \frac{u(x_0, y_0 + t) - u(x_0, y_0) + i(v(x_0, y_0 + t) - v(x_0, y_0))}{it}$$
$$\to f'(z_0),$$

as $t \to 0$, with $t \neq 0$. Again, this means that the real and imaginary parts of the left hand side separately have limits. Hence the partial derivatives u_y and v_y exist at (x_0, y_0) and

$$-iu_y(x_0, y_0) + v_y(x_0, y_0) = f'(z_0).$$

Equating these two expressions for $f'(z_0)$ leads to the following relations.

$$\left. \begin{array}{l} u_x(x_0, y_0) = v_y(x_0, y_0) \\ u_y(x_0, y_0) = -v_x(x_0, y_0) \end{array} \right\} \quad \textbf{Cauchy-Riemann equations}$$

If we denote complex differentiation by D_z, then the discussion above tells us that

$$D_z = \partial_x = -i\partial_y$$

where it is understood that D_z is applied to $f(z)$ and the partial derivatives are applied to $u(x, y) + iv(x, y)$. Thus

$$\partial_x u + i\,\partial_x v = D_z f = f'$$
$$= -i\,\partial_y(u + iv) = \partial_y v - i\,\partial_y u.$$

Equating real and imaginary parts gives the Cauchy-Riemann equations. We have proved the following theorem.

Theorem 4.1 *If the function f is differentiable at $z_0 = x_0 + iy_0$ then f satisfies the Cauchy-Riemann equations at (x_0, y_0).*

Remark 4.3 If f does *not* satisfy the Cauchy-Riemann equations at some point (x_0, y_0), then certainly f fails to be differentiable at $z_0 = x_0 + iy_0$.

Example 4.2 We have already seen that the function $f(z) = \overline{z}$ is nowhere differentiable. Let us consider the Cauchy-Riemann equations for this f.

We have $f(z) = \overline{z} = x - iy$ and so $u(x, y) = x$ and $v(x, y) = -y$. Evidently both u and v possess partial derivatives everywhere in $\mathbb{R}^2$, but $u_x = 1$, $u_y = 0$, $v_x = 0$ and $v_y = -1$. We see that u_x is never equal to v_y and so the Cauchy-Riemann equations are never valid. We can therefore conclude that f is nowhere differentiable (as we already knew).

Remark 4.4 Is the converse of this last theorem true, that is, if f satisfies the Cauchy-Riemann equations at some point (x_0, y_0), is it true that f is differentiable at $z = x_0 + iy_0$? The answer, in general, is no!

Example 4.3 Let f be the function $f(x + iy) = |xy|^{1/2}$. Then we see that $u(x, y) = |xy|^{1/2}$, $v(x, y) = 0$, and

$$u_x(0, 0) = u_y(0, 0) = 0 = v_x(0, 0) = v_y(0, 0).$$

(For example, $u_x(0, 0) = \lim_{s \to 0} \dfrac{u(s, 0) - u(0, 0)}{s} = 0$.) From this, it follows that the Cauchy-Riemann equations hold for f at $(0, 0)$.

However, consider $s + is$ as $s \to 0$. We have, for $s \neq 0$,

$$\frac{f(s + is) - f(0)}{s + is} = \frac{|s|}{s + is} = \frac{|s|(s - is)}{2s^2} = \frac{|s|(1 - i)}{2s}$$

$$\to \begin{cases} \frac{1}{2}(1 - i), & \text{if } s \to 0 \text{ through positive values,} \\ -\frac{1}{2}(1 - i), & \text{if } s \to 0 \text{ through negative values.} \end{cases}$$

We conclude that f is not differentiable at $z = 0$ even though f satisfies the Cauchy-Riemann equations there.

Under continuity conditions on the partial derivatives, we can prove the converse. We shall use the following result from the theory of real functions of two real variables.

Lemma 4.1   *Suppose that $\psi(x, y)$ is defined in some disc around the point $(x_0, y_0) \in \mathbb{R}^2$ and has continuous partial derivatives in this disc. Then, for all sufficiently small h and k in $\mathbb{R}$*

$$\psi(x_0 + h, y_0 + k) - \psi(x_0, y_0) = h\,\psi_x(x_0, y_0) + k\,\psi_y(x_0, y_0) + R,$$

where $|R| / \sqrt{h^2 + k^2} \to 0$ as $h, k \to 0$ (with $h^2 + k^2 \neq 0$).

Proof.   The idea is to find an expression for R and then use the Mean Value Theorem to obtain the required estimate. Indeed, we have

$$\psi(x_0 + h, y_0 + k) - \psi(x_0, y_0)$$
$$= \psi(x_0 + h, y_0 + k) - \psi(x_0 + h, y_0) + \psi(x_0 + h, y_0) - \psi(x_0, y_0)$$
$$= k\,\psi_y(x_0 + h, y_0 + \theta k) + \psi(x_0 + h, y_0) - \psi(x_0, y_0)$$

for some $0 \leq \theta \leq 1$, by the Mean Value Theorem applied to the function $y \mapsto \psi(x_0 + h, y)$ for y between y_0 and $y_0 + k$,

$$= h\,\psi_x(x_0, y_0) + k\,\psi_y(x_0, y_0) + R$$

where $R = \alpha_1 + \alpha_2$ with

$$\alpha_1 = k\,(\psi_y(x_0 + h, y_0 + \theta k) - \psi_y(x_0, y_0))$$

and

$$\alpha_2 = \psi(x_0 + h, y_0) - \psi(x_0, y_0) - h\,\psi_x(x_0, y_0).$$

Now, if $h, k \to 0$, with $h^2 + k^2 \neq 0$, then

$$\left| \frac{\alpha_1}{\sqrt{h^2 + k^2}} \right| = \underbrace{\left| \frac{k}{\sqrt{h^2 + k^2}} \right|}_{\leq 1} \underbrace{\left| \psi_y(x_0 + h, y_0 + \theta k) - \psi_y(x_0, y_0) \right|}_{\to 0}$$

by continuity of ψ_y. Also $\alpha_2/\sqrt{h^2 + k^2} = 0$ if $h = 0$, otherwise (i.e., $h \neq 0$)

$$\left| \frac{\alpha_2}{\sqrt{h^2 + k^2}} \right| = \underbrace{\left| \frac{h}{\sqrt{h^2 + k^2}} \right|}_{\leq 1} \left| \underbrace{\frac{\psi(x_0 + h, y_0) - \psi(x_0, y_0)}{h}}_{\to \psi_x(x_0, y_0)} - \psi_x(x_0, y_0) \right|$$

$$\to 0$$

as $h, k \to 0$, with $h \neq 0$. The result follows. □

Theorem 4.2 *Let $f = u + iv$ and suppose that the partial derivatives u_x, u_y, v_x and v_y exist for all (x, y) with $x + iy$ in some open disc around $x_0 + iy_0$ and that these partial derivatives are continuous at (x_0, y_0). Suppose further, that f satisfies the Cauchy-Riemann equations at (x_0, y_0). Then f is complex differentiable at the point $z_0 = x_0 + iy_0$.*

Proof. Let $\lambda = u_x(x_0, y_0) = v_y(x_0, y_0)$ and $\mu = -u_y(x_0, y_0) = v_x(x_0, y_0)$. By lemma 4.1, we can write

$$u(x_0 + h, y_0 + k) - u(x_0, y_0) = h\lambda - k\mu + R_1$$

and

$$v(x_0 + h, y_0 + k) - v(x_0, y_0) = h\mu + k\lambda + R_2$$

for small h, k and where $R_1/\sqrt{h^2 + k^2}$ and $R_2/\sqrt{h^2 + k^2}$ tend to zero as $h, k \to 0$ (not both zero). Hence, with $\zeta = h + ik \neq 0$,

$$\begin{aligned} \frac{f(z_0 + \zeta) - f(z_0)}{\zeta} &= \frac{h\lambda - k\mu + i(h\mu + k\lambda) + R_1 + iR_2}{h + ik} \\ &= \frac{(h + ik)(\lambda + i\mu) + R_1 + iR_2}{h + ik} \\ &= \lambda + i\mu + \underbrace{\frac{R_1 + iR_2}{h + ik}}_{\to 0 \text{ as } h + ik \to 0}. \end{aligned}$$

It follows that f is differentiable at $z_0 = x_0 + iy_0$ and that its derivative is given by $f'(z_0) = \lambda + i\mu$. □

4.5   Analytic Functions

Definition 4.3   Let D be a domain and $f : D \to \mathbb{C}$. The function f is
said to be analytic at the point $z_0 \in D$ if and only if there is some $r > 0$
such that f is differentiable at every point in the disc $D(r, z_0)$.

   If f is analytic at every point of D, we say that f is analytic in D. The
set of functions analytic in a domain D is denoted $H(D)$. If the function
$f : \mathbb{C} \to \mathbb{C}$ is analytic at every point in $\mathbb{C}$, then we say that f is entire.

Remark 4.5   Analytic functions are also called holomorphic functions.
Note that to say that a function f is analytic at every point of an open set
G is the same as saying that f is differentiable at every point in G. (This
is because every point in G is the centre of a disc lying in G.)

Examples 4.4

(1) For any $n \in \mathbb{N}$, the function $f(z) = z^n$, $z \in \mathbb{C}$, is entire. The function
    $g(z) = z^{-n}$, $z \neq 0$, $n \in \mathbb{N}$ is analytic in the punctured plane, $\mathbb{C} \setminus \{0\}$.
(2) For any fixed $\zeta \in \mathbb{C}$, the function $f(z) = 1/(z-\zeta)$ is analytic in $\mathbb{C} \setminus \{\zeta\}$.

Proposition 4.6   *Suppose the function f is real-valued. Then either f is
not differentiable at z_0 or $f'(z_0) = 0$.*

Proof.   Suppose that f is differentiable at z_0. Then

$$\lim_{z \to z_0} \frac{f(z) - f(z_0)}{z - z_0} = f'(z_0).$$

Set $\zeta = s + it \neq 0$ and let $z = z_0 + \zeta$.

   Taking $t = 0$, we have $\dfrac{f(z_0 + s) - f(z_0)}{s} \to f'(z_0)$ as $s \to 0$. But the
left hand side is real-valued, and so we must have that $f'(z_0) \in \mathbb{R}$.

   Now, take $s = 0$. Then $\dfrac{f(z_0 + it) - f(z_0)}{it} \to f'(z_0)$. However, the left
hand side is purely imaginary and so $if'(z_0) \in \mathbb{R}$. This is only possible if
$f'(z_0) = 0$.   □

Theorem 4.3   *Let $f \in H(D)$ and suppose that $f'(z) = 0$ for every $z \in D$.
Then f is constant on D.*

Proof.   Suppose first that $[z', z''] \subset D$ where $z' \neq z''$. For $0 \le t \le 1$ put
$z(t) = z' + t(z'' - z')$, and let $\varphi = \operatorname{Re} f$ and $\psi = \operatorname{Im} f$.

Fix $t_0 \in (0,1)$. Then $t \neq t_0$ implies that $z(t) \neq z(t_0)$ and

$$\frac{f(z(t)) - f(z(t_0))}{t - t_0} = \left(\frac{f(z(t)) - f(z(t_0))}{z(t) - z(t_0)}\right)\left(\frac{z(t) - z(t_0)}{t - t_0}\right)$$
$$\to f'(z(t_0))\,(z'' - z') = 0, \text{ by hypothesis,}$$

as $t \to t_0$. Taking real and imaginary parts of the left hand side, we conclude that

$$\frac{\varphi(z(t)) - \varphi(z(t_0))}{t - t_0} \to 0$$

and

$$\frac{\psi(z(t)) - \psi(z(t_0))}{t - t_0} \to 0$$

as $t \to t_0$. If we set $\alpha(t) = \varphi(z(t))$ and $\beta(t) = \psi(z(t))$, then we have shown that the real functions α and β of the real variable t are differentiable at each $t_0 \in (0,1)$ with $\alpha'(t_0) = \beta'(t_0) = 0$. It follows, by real analysis, that α and β are both constant on $(0,1)$. By continuity, they are both constant on the interval $[0,1]$. In particular,

$$\begin{aligned}
f(z') &= \varphi(z') + i\psi(z') \\
&= \alpha(0) + i\beta(0) \\
&= \alpha(1) + i\beta(1) \\
&= \varphi(z'') + i\psi(z'') \\
&= f(z'').
\end{aligned}$$

Now let ζ and ξ be any pair of points in D. Since D is connected, there is a polygon with vertices $\zeta = z_0, z_1, \ldots, z_n = \xi$ joining ζ to ξ in D. By the above argument, applied to the line segments $[z_0, z_1], \ldots, [z_{n-1}, z_n]$, one by one, we get

$$f(\zeta) = f(z_1) = \cdots = f(z_{n-1}) = f(\xi).$$

It follows that f is constant on D. $\qquad\square$

Remark 4.6 Another proof can be given using the Cauchy-Riemann equations, as follows. We have seen that

$$f'(x + iy) = \varphi_x(x,y) + i\psi_x(x,y) = -i\varphi_y(x,y) + \psi_y(x,y)\,.$$

But $f' = 0$ on D, by hypothesis, and therefore $\varphi_x = \varphi_y = \psi_x = \psi_y = 0$ throughout D. Suppose that the line segment $[x_0+iy_0, x_0+h+iy_0]$ lies in D. Applying the Mean Value Theorem to the real-valued function $x \mapsto \varphi(x, y_0)$ for $x \in [x_0, x_0 + h]$, we see that $\varphi(x_0, y_0) = \varphi(x_0 + h, y_0)$. In other words, φ has the same value at each end of the (horizontal) line segment. The same argument applies to ψ and similar reasoning shows that this is also true for (vertical) line segments of the form $[x_0 + iy_0, x_0 + i(y_0 + k)]$.

Now let ζ and ξ be any two points in D. Since D is open and connected, it is stepwise connected (by theorem 3.7) and so ζ and ξ can be joined by a step-path. Applying the above discussion to each of the line segments making up such a path, we deduce that φ has the same value at the start and end of the whole path (in fact, is constant throughout). The same is true of ψ and so $f = \varphi + i\psi$ has the same value at each end, that is, $f(\zeta) = f(\xi)$.

Corollary 4.1   *Suppose that $f \in H(D)$ and f is real-valued. Then f is constant on D.*

Proof.   Since $f \in H(D)$, we know that f is differentiable at every $z \in D$. But then the fact that f is real-valued means that we must have $f'(z) = 0$, for all $z \in D$. Hence, by the theorem, f is constant on D.   □

Corollary 4.2   *Suppose that $f \in H(D)$ and that $|f|$ is constant on D. Then f is constant on D.*

Proof.   Suppose that $|f(z)|^2 = \alpha$ for all $z \in D$. If $\alpha = 0$, then f vanishes on D and we are done. So suppose that $\alpha \neq 0$. Then f is never zero on D and so $1/f \in H(D)$. But $\alpha = |f|^2 = \overline{f} f$, which means that $\overline{f} = \alpha/f$ is analytic in D. Hence the real-valued function $f + \overline{f}$ belongs to $H(D)$ and so it is constant. Similarly, $i(f - \overline{f})$ belongs to $H(D)$ and, since it is real-valued, it must also be constant. Therefore

$$f = \tfrac{1}{2}\big((f + \overline{f}) - i(i(f - \overline{f})) \big)$$

is constant on D.   □

Example 4.5   Let D be a domain and let $f \in H(D)$. Suppose that there is some straight line L in $\mathbb{C}$ such that $f(z) \in L$ for every $z \in D$. Then f is constant on D. Indeed, any straight line L in $\mathbb{C}$ has the form $L = \{ z : z = z_0 + t\zeta,\ t \in \mathbb{R} \}$ for suitable z_0 and $\zeta \neq 0$. If $f(z) \in L$, then $(f(z) - z_0)/\zeta \in \mathbb{R}$ for all $z \in D$ and so is constant (because it is analytic and real-valued on D). But this means that f is also constant on D.

4.6 Power Series

A complex power series is a series of the form $\sum_{n=0}^{\infty} a_n(z - z_0)^n$, with z, z_0 and $a_n \in \mathbb{C}$ (it is a series of "powers", $(z - z_0)^n$). The absolute convergence of such a power series is, by definition, determined by the convergence of the real power series $\sum_{n=0}^{\infty} |a_n| |w|^n$, where we have set $w = (z - z_0)$.

Clearly, the power series converges (also absolutely) when $z = z_0$. If we let S denote those points $z \in \mathbb{C}$ for which this power series converges, then it is natural to ask what S can look like. Evidently, $z_0 \in S$ and so S is not empty. Suppose that $\zeta \in S$ with $\zeta \neq z_0$. Then $\sum_{n=0}^{\infty} a_n(\zeta - z_0)^n$ converges and so $|a_n(\zeta - z_0)^n| \to 0$ as $n \to \infty$. In particular, the collection $\{ |a_n(\zeta - z_0)^n| : n \in \mathbb{N} \}$ is bounded, i.e., there is some $M > 0$ such that $|a_n| \rho^n \leq M$ for all $n \in \mathbb{N}$, where we have set $\rho = |\zeta - z_0|$. For any point z in the disc $D(z_0, \rho)$, we see that

$$|a_n(z - z_0)^n| = |a_n| \rho^n \left(\frac{|(z - z_0)|}{\rho} \right)^n \leq M r^n$$

where $r = |z - z_0| / \rho < 1$. It follows (by the Comparison Test) that the power series converges *absolutely* for all z in the open disc $D(z_0, \rho)$. If S contains any point $\zeta \neq z_0$, then S must also contain the whole open disc $D(z_0, |\zeta - z_0|)$ (and in this open disc, the power series converges absolutely).

If the set S is unbounded, then clearly, the power series will converge (and also absolutely) for all $z \in \mathbb{C}$. On the other hand, if S is bounded, it could consist of just the single point z_0 or it could contain other points. In this second case, there will be some $\hat{R} > 0$ such that the power series converges absolutely for all z in the open disc $D(z_0, \hat{R})$ but diverges at every point z outside the closed disc $\overline{D(z_0, \hat{R})}$. Indeed, $\hat{R}$ is given by $\sup\{ |z - z_0| : z \in S \}$. The discussion so far says nothing at all about the behaviour of the power series *on* the circle $|z - z_0| = \hat{R}$. This will depend very much on the details of the series in question and will vary from power series to power series.

Examples 4.6

(1) The power series $\sum_{n=0}^{\infty} z^n$ converges (absolutely) for all z with $|z| < 1$. It diverges for all other values of z. In particular, it diverges on the circle $|z| = 1$ (since then z^n does not converge to zero).

(2) The power series $\sum_{n=1}^{\infty} z^n/n^2$ converges (absolutely) for z with $|z| \leq 1$ but diverges for all other z (since z^n/n^2 does not converge to 0 when $|z| > 1$).

(3) The power series $\sum_{n=1}^{\infty} z^n/n$ converges (absolutely) for all z obeying $|z| < 1$. It diverges for $z = 1$ (where it becomes the divergent harmonic series $1 + \frac{1}{2} + \frac{1}{3} + \frac{1}{4} + \dots$) and so it necessarily diverges for all z with $|z| > 1$. One can show that it converges (but *not* absolutely) for all z with $|z| = 1$ *except* for the point $z = 1$. (For fixed $z \neq 1$ with $|z| = 1$, the (moduli of the) partial sums $\sigma_k = z + z^2 + \cdots + z^k = (z - z^{k+1})/(1 - z)$ are bounded (by $2/(|1 - z|)$). An application of Dirichlet's Test (applied separately to the real and imaginary parts) gives the stated convergence.)

The value of $\hat{R}$ introduced above is called the radius of convergence of the power series $\sum_{n=0}^{\infty} a_n (z - z_0)^n$; where we say that $\hat{R} = 0$ if the series converges absolutely only for $z = z_0$, i.e., only for $w = 0$, and that the series has an infinite radius of convergence if it converges absolutely for all values of $z - z_0$, i.e., for all values of z. The disc $D(z_0, \hat{R})$ is called the disc of convergence of the power series.

4.7 The Derived Series

The (complex) derivative of the typical term $a_n (z - z_0)^n$ in the power series is $n a_n (z - z_0)^{n-1}$. This leads to a new power series, called the derived series. We wish to show that a power series is differentiable everywhere inside its disc of convergence, and, moreover, that its derivative is got by simply differentiating term by term.

To avoid notational complications, we shall consider the case $z_0 = 0$ and then apply the chain rule to recover the general situation. First, however, we must establish convergence of the derived series.

Proposition 4.7 *Suppose the power series $f(z) = \sum_{n=0}^{\infty} a_n z^n$ converges absolutely for $|z| < R$. Then the derived series $g(z) = \sum_{n=1}^{\infty} n a_n z^{n-1}$ also converges absolutely for $|z| < R$.*

Proof. Let z with $|z| < R$ be given and let r satisfy $|z| < r < R$. Then $\sum_{n=0}^{\infty} |a_n| r^n$ is convergent and so certainly there is some constant $K > 0$ such that $|a_n| r^n \leq K$ for all n (the terms of a convergent series actually converge to zero, so the sequence of terms is bounded). Hence

$$\left| n a_n z^{n-1} \right| = n |a_n| r^{n-1} \left| \frac{z}{r} \right|^{n-1} \leq n \frac{K}{r} \left| \frac{z}{r} \right|^{n-1}.$$

But if we set $t = |z/r|$, then $t < 1$ and so the series $\sum_{n=1}^{\infty} n t^{n-1}$ converges (to $1/(1-t)^2$). It follows, by the Comparison Test, that the derived series g converges absolutely, as claimed. □

Now let us tackle the question of the differentiability of f.

Theorem 4.4 *Suppose the power series $f(z) = \sum_{n=0}^{\infty} a_n z^n$ converges absolutely for $|z| < R$. Then f is differentiable at any z with $|z| < R$, and its derivative, $f'(z)$, is given by the absolutely convergent power series $\sum_{n=1}^{\infty} n\, a_n z^{n-1}$. In other words, the derivative of a power series is its derived series (inside its disc of convergence).*

Proof. We have seen that the power series $g(z) \equiv \sum_{n=1}^{\infty} n\, a_n z^{n-1}$ converges absolutely. We must show that $(f(w)-f(z))/(w-z)-g(z)$ converges to zero, as $w \to z$. To do this, we use the fact that convergent series can be manipulated termwise and so this expression can be written also as a series. Next, we split this into two parts and estimate each one separately.

Our first observation, then, is that for $w \neq z$

$$\frac{f(w) - f(z)}{w - z} - g(z) = \sum_{n=1}^{\infty} a_n \frac{w^n - z^n}{w - z} - n a_n z^{n-1}$$

$$= \sum_{n=1}^{\infty} a_n \left(w^{n-1} + w^{n-2}z + w^{n-3}z^2 + \cdots \right.$$

$$\left. \cdots + wz^{n-2} + z^{n-1} - nz^{n-1}\right)$$

$$= \phi_1(w) + \phi_2(w) \tag{$*$}$$

where

$$\phi_1(w) = \sum_{n=1}^{N} a_n \left(w^{n-1} + w^{n-2}z + w^{n-3}z^2 + \cdots + wz^{n-2} + z^{n-1} - nz^{n-1}\right)$$

and

$$\phi_2(w) = \sum_{n=N+1}^{\infty} a_n \left(w^{n-1} + w^{n-2}z + w^{n-3}z^2 + \cdots \right.$$

$$\left. \cdots + wz^{n-2} + z^{n-1} - nz^{n-1}\right).$$

We will say something about N shortly. Let $|z| < R$ and $\varepsilon > 0$ be given and let r satisfy $|z| < r < R$. Then, for any w with $|w| < r$, we have

$$\left|a_n\left(w^{n-1} + w^{n-2}z + \ldots + wz^{n-2} + z^{n-1} - nz^{n-1}\right)\right|$$
$$\leq |a_n|\left(\left|w^{n-1}\right| + \left|w^{n-2}z\right| + \ldots\right.$$
$$\left.\ldots + \left|wz^{n-2}\right| + \left|z^{n-1}\right| + n\left|z^{n-1}\right|\right)$$
$$\leq |a_n|\, 2nr^{n-1}.$$

The series $\sum_{n=1}^{\infty} n\,|a_n|\,r^{n-1}$ converges (the series for g converges absolutely for $|z| = r$) and so we may choose N sufficiently large that $|\phi_2(w)|$, the modulus of the second term on the right hand side of $(*)$ (the tail), is less than $\varepsilon/2$ (it is bounded by the tail of a convergent series). Fix N so that this is so.

Next, we consider $\phi_1(w)$, the first term on the right hand side of $(*)$. This is a sum of N terms, each of which tends to zero as $w \to z$. It follows that there is $\delta' > 0$ such that $|\phi_1(w)| < \varepsilon/2$, whenever $|w - z| < \delta'$. Now we piece these two arguments together.

Let $\delta = \min\{\delta', r - |z|\}$. Then, if $0 < |w - z| < \delta$, we have that $0 < |w - z| < \delta'$ and also that $|w| \leq |w - z| + |z| < (r - |z|) + |z| = r$. Therefore

$$\left|\frac{f(w) - f(z)}{w - z} - g(z)\right| = |\phi_1(w) + \phi_2(w)|$$
$$\leq |\phi_1(w)| + |\phi_2(w)|$$
$$< \tfrac{1}{2}\varepsilon + \tfrac{1}{2}\varepsilon = \varepsilon,$$

whenever $0 < |w - z| < \delta$, and the proof is complete. $\qquad\square$

Corollary 4.3 *Suppose that the power series* $f(z) = \sum_{n=0}^{\infty} a_n(z - z_0)^n$ *converges absolutely for* $|z - z_0| < R$. *Then* f *is differentiable at each* $z \in D(z_0, R)$ *with derivative* $f'(z) = \sum_{n=1}^{\infty} n\,a_n(z - z_0)^{n-1}$.

Proof. Let $h(w) = \sum_{n=0}^{\infty} a_n w^n$. By hypothesis, the power series for $h(w)$ converges absolutely for all $|w| < R$. In particular, h is differentiable with derivative $h'(w) = \sum_{n=1}^{\infty} n a_n w^{n-1}$, for $|w| < R$. Let $\psi(z) = z - z_0$. Then ψ is differentiable and $\psi'(z) = 1$ for all z. By the chain rule, $h \circ \psi = h(\psi(z))$ is differentiable for z with $|\psi(z)| < R$ and, for such z, its derivative is given by $(h \circ \psi)'(z) = h'(\psi(z))\psi'(z)$, as required. $\qquad\square$

Corollary 4.4 *Suppose that the power series* $f(z) = \sum_{n=0}^{\infty} a_n(z - z_0)^n$ *converges absolutely for* $z \in D(z_0, R)$. *Then for any* $k \in \mathbb{N}$, f *is* k-*times* *differentiable with k^{th}-derivative given by*

$$f^{(k)}(z) = \sum_{n=k}^{\infty} a_n\, n(n-1)\ldots(n-(k-1))(z-z_0)^{n-k},$$

where this last series converges absolutely for $z \in D(z_0, R)$. In particular,

$$f^{(k)}(z_0) = k!\, a_k.$$

Proof. The proof is by induction on k. We know, by corollary 4.3, that the result is true for $k = 1$. Suppose it is true for $k = m$. Write

$$g(z) = f^{(m)}(z) = \sum_{n=m}^{\infty} a_n\, n(n-1)\ldots(n-(m-1))(z-z_0)^{n-m}$$

$$= \sum_{j=0}^{\infty} a_{m+j}\, (m+j)(m+j-1)\ldots(j+1)(z-z_0)^j.$$

By corollary 4.3, g is differentiable at $z \in D(z_0, R)$ with derivative given by the absolutely convergent power series

$$g'(z) = \sum_{j=1}^{\infty} a_{m+j}\, (m+j)(m+j-1)\ldots(j+1)j\, (z-z_0)^{j-1}.$$

Relabelling, the result follows for $k = m+1$, and, by induction, the proof of the formula for $f^{(k)}(z)$ is complete. Setting $z = z_0$ completes the proof since only the first term in the series for $f^{(k)}(z_0)$ survives. $\qquad\square$

What's going on? These results tell us that power series behave very much like polynomials—as long as we stay inside their discs of convergence. The behaviour on the boundary of these discs varies from power series to power series and can be very complicated.

4.8 Identity Theorem for Power Series

We will need the following result later on.

Theorem 4.5 (Identity Theorem for Power Series) *Suppose that the power series $f(z) = \sum_{n=0}^{\infty} a_n(z-z_0)^n$ and $g(z) = \sum_{n=0}^{\infty} b_n(z-z_0)^n$ both converge absolutely for all $z \in D(z_0, R)$. Suppose, further, that there is some sequence (ζ_k) in $D(z_0, R)$, with $\zeta_k \neq z_0$ for all k, such that $\zeta_k \to z_0$ as $k \to \infty$, and such that $f(\zeta_k) = g(\zeta_k)$ for all k. Then $a_n = b_n$ for all n, that is, $f = g$.*

Proof. Write $h(z) = \sum_{n=0}^{\infty} c_n(z - z_0)^n$, where $c_n = a_n - b_n$. Then $h(\zeta_k) = 0$ for all k. Suppose that c_m is the first non-zero coefficient of h, i.e., $c_n = 0$ for $n < m$ and $c_m \neq 0$. In this case, we can write h as

$$h(z) = (z - z_0)^m \underbrace{\left(c_m + c_{m+1}(z - z_0) + \cdots \right)}_{\varphi(z)}.$$

Now, $h(\zeta_k) = 0$ but $\zeta_k \neq z_0$ and so we must have that $\varphi(\zeta_k) = 0$, for all k. But φ is a power series which converges absolutely for z in the disc $D(z_0, R)$. In particular, φ is continuous at z_0. Hence $\varphi(\zeta_k) \to \varphi(z_0)$ and we conclude that $\varphi(z_0) = 0$. This means that $c_m = 0$, which is a contradiction. It follows that every c_n vanishes and so $a_n = b_n$, for all n, and $f = g$. $\square$

What's going on ?   The theorem above simply amounts to the statement that if the centre of the disc of convergence of a power series is a limit point of zeros, then all the coefficients of the power series are zero. The series is identically zero. It is the zero power series.

Chapter 5

The Complex Exponential and Trigonometric Functions

5.1 The Functions $\exp z$, $\sin z$ and $\cos z$

We take, as our starting point, the definitions of the exponential and the trigonometric functions as complex power series. We will see that they have the expected properties.

Definition 5.1 The complex exponential function, $\exp z$, the complex sine function, $\sin z$, and the complex cosine function, $\cos z$, are defined by the power series as follows:

$$\exp z - \sum_{n=0}^{\infty} \frac{z^n}{n!} = 1 + z + \frac{z^2}{2!} + \frac{z^3}{3!} + \frac{z^4}{4!} + \ldots$$

$$\sin z = \sum_{n=0}^{\infty} \frac{(-1)^n z^{2n+1}}{(2n+1)!} = z - \frac{z^3}{3!} + \frac{z^5}{5!} - \frac{z^7}{7!} + \ldots$$

$$\cos z = \sum_{n=0}^{\infty} \frac{(-1)^n z^{2n}}{(2n)!} = 1 - \frac{z^2}{2!} + \frac{z^4}{4!} - \frac{z^6}{6!} + \ldots .$$

The Ratio Test shows that each of these series is absolutely convergent for all $z \in \mathbb{C}$, that is, they each have infinite radius of convergence. Thus, the functions exp, sin and cos are entire functions. If z is real, we recover precisely the real series expressions for these functions.

One readily calculates the various derived series, and the result is that

$$\exp' z = \exp z$$
$$\sin' z = \cos z$$
$$\cos' z = -\sin z$$

for all $z \in \mathbb{C}$. Evidently, $\exp 0 = 1$, $\sin 0 = 0$ and $\cos 0 = 1$. Furthermore, sin is an odd function and cos is an even function, that is, $\sin(-z) = -\sin z$, and $\cos(-z) = \cos z$, for all $z \in \mathbb{C}$. If x is real, then each of $\exp x$, $\sin x$ and $\cos x$ is also real. The relationship between these three functions is not particularly transparent in the real context, but direct substitution shows that, for any $z \in \mathbb{C}$,

$$\exp iz = \cos z + i \sin z.$$

From this it follows that

$$\cos z = \frac{\exp iz + \exp(-iz)}{2} \quad \text{and} \quad \sin z = \frac{\exp iz - \exp(-iz)}{2i}.$$

5.2 Complex Hyperbolic Functions

Definition 5.2 The complex hyperbolic sine and cosine functions are defined as the series

$$\sinh z = \sum_{n=0}^{\infty} \frac{z^{2n+1}}{(2n+1)!} = z + \frac{z^3}{3!} + \frac{z^5}{5!} + \frac{z^7}{7!} + \cdots$$

$$\cosh z = \sum_{n=0}^{\infty} \frac{z^{2n}}{(2n)!} = 1 + \frac{z^2}{2!} + \frac{z^4}{4!} + \frac{z^6}{6!} + \cdots.$$

We see that

$$\sinh z = \tfrac{1}{2}(\exp z - \exp(-z)) \quad \text{and} \quad \cosh z = \tfrac{1}{2}(\exp z + \exp(-z))$$

and, moreover, that $\sinh z = -i \sin(iz)$ and $\cosh z = \cos(iz)$. Thus, in the complex variable context, the properties of the hyperbolic functions can be readily obtained from those of the complex trigonometric functions, which shows that this is really the natural context for these functions. This is in sharp contrast to the real variable situation, where the behaviour of the hyperbolic functions on the one hand, and that of the real trigonometric functions on the other, are quite distinct. It is the extension from a real to a complex variable that exposes the otherwise hidden connections.

5.3 Properties of exp z

The basic properties of the complex exponential function are given in the following proposition.

Proposition 5.1   *The exponential function has the following properties:*

  (i) $\exp 0 = 1$,

  (ii) $\exp(z + w) = \exp z \, \exp w$, *for any* $z, w \in \mathbb{C}$,

 (iii) $\exp z \neq 0$, *for all* $z \in \mathbb{C}$,

 (iv) $\exp(-z) = 1/\exp z$, *for all* $z \in \mathbb{C}$,

  (v) $\exp(x + iy) = \exp x \, (\cos y + i \sin y)$, *for* $x, y \in \mathbb{C}$ *(and, in particular, for any* $x, y \in \mathbb{R}$*).*

Proof.   Setting $z = 0$ immediately gives (i). To prove (ii), fix $w \in \mathbb{C}$ and set $f(z) = \exp(z + w) \exp(-z)$. Then we find that, for any $z \in \mathbb{C}$,

$$
\begin{aligned}
f'(z) &= \big(\exp(z + w)\big)' \exp(-z) + \exp(z + w) \big(\exp(-z)\big)' \\
&= \exp(z + w) \exp(-z) - \exp(z + w) \exp(-z) \\
&= 0.
\end{aligned}
$$

It follows that f is constant, and therefore $f(z) = f(0) = \exp w$, that is,

$$
\exp w = \exp(z + w) \, \exp(-z).
$$

Now set $w = a + b$ and $z = -b$. Then we obtain

$$
\exp(a + b) = \exp a \, \exp b,
$$

for any $a, b \in \mathbb{C}$, as required.

Using (ii) with $w = -z$, we find that

$$
\exp z \, \exp(-z) = \exp 0 = 1, \ \text{ by (i),}
$$

and so (iii) follows, and so does (iv).

To prove (v), let $z = x + iy$ and then apply (ii) to obtain

$$
\begin{aligned}
\exp z = \exp(x + iy) &= \exp x \, \exp(iy) \\
&= \exp x \, (\cos y + i \sin y).
\end{aligned}
$$

This holds for any $x, y \in \mathbb{C}$ and so, in particular, also for real x, y. $\square$

Corollary 5.1   *Let e be the real number given by*

$$
e = 1 + 1 + \frac{1}{2!} + \frac{1}{3!} + \frac{1}{4!} + \ldots = \exp 1.
$$

Then $\exp n = e^n$ for any $n \in \mathbb{Z}$.

Proof. For each $n \in \mathbb{Z}$, let $P(n)$ be the statement that $\exp n = e^n$. For $n \in \mathbb{N}$, we shall prove the claim by induction. By definition, $\exp 1 = e$ and so $P(1)$ is true. Now let $n \in \mathbb{N}$ and suppose that $P(n)$ is true. Then

$$\exp(n+1) = \exp n \, \exp 1, \text{ by the previous proposition,}$$
$$= e^n \exp 1, \text{ by induction hypothesis,}$$
$$= e^n e, \text{ since } \exp 1 = e,$$
$$= e^{n+1}$$

so that $P(n+1)$ is true. By induction, $P(n)$ is true for all $n \in \mathbb{N}$.

Clearly $P(0)$ is true, since both $\exp 0$ and e^0 are equal to 1 ($e^0 = 1$, by definition). Now let $n = -m$ with $m \in \mathbb{N}$. Then $\exp m = e^m$ and

$$\exp n = \exp(-m) = \frac{1}{\exp m} = \frac{1}{e^m} = e^{-m} = e^n,$$

as required. $\square$

Remark 5.1 It is because of this relationship that one writes e^z for $\exp z$. This notation is often very convenient.

Remark 5.2 Suppose f is entire and satisfies $f'(z) = f(z)$ for all z and $f(0) = 1$. Then $f(z) = \exp z$. To see this, we consider the entire function $g(z) = f(z) \exp(-z)$. We see that $g'(z) = 0$ for all $z \in \mathbb{C}$ and so g is constant, $g(z) = g(0) = 1$. But then we find that $f(z) = \exp z$, as required.

For $z = x + iy$, let $f(z) = e^x(\cos y + i \sin y)$ so that $\operatorname{Re} f(x + iy) = e^x \cos y$ and $\operatorname{Im} f(x + iy) = e^x \sin y$. These functions have continuous partial derivatives throughout $\mathbb{R}^2$ and obey the Cauchy-Riemann equations and so $f(z)$ is analytic at every $z \in \mathbb{C}$. Furthermore, we know that its derivative is given by

$$f'(x+iy) = \partial_x \operatorname{Re} f(x+iy) + i\partial_x \operatorname{Im} f(x+iy) = e^x \cos y + ie^x \sin y = f(x+iy).$$

Since $f(0) = 1$ we conclude that $f(z) = \exp z$.

Sometimes this approach is used to *define* the complex exponential function—assuming that the real functions e^t, $\sin \theta$ and $\cos \theta$ are somehow already known.

Proposition 5.2 *The number e is irrational.*

Proof. For any $k \in \mathbb{N}$, the power series expression for $e = \exp 1$ implies that

$$\sum_{j=0}^{k} \frac{1}{j!} < e = \sum_{j=0}^{k} \frac{1}{j!} + \frac{1}{k!}\left(\frac{1}{k+1} + \frac{1}{(k+1)(k+2)} + \cdots\right).$$

Since $1/(k+1)(k+2)\ldots(k+m) < 1/2^m$ for any $m \in \mathbb{N}$, $m > 1$, we see that the term in brackets on the right hand side above is bounded above by $\sum_{m=1}^{\infty} 1/2^m = 1$ and so

$$\sum_{j=0}^{k} \frac{1}{j!} < e < \sum_{j=0}^{k} \frac{1}{j!} + \frac{1}{k!}$$

for any $k \in \mathbb{N}$. Multiplying through by $k!$ and rearranging, we get

$$0 < k!\left(e - \sum_{j=0}^{k} \frac{1}{j!}\right) < 1$$

for any $k \in \mathbb{N}$. If e were rational, then we could write e as $e = p/q$ for some $p, q \in \mathbb{N}$. Setting k in the inequality above to be any integer greater than q and then taking $k!$ inside the bracket, we see that the middle expression is a positive integer strictly less than 1, which is impossible. We conclude that e is irrational. □

Remark 5.3 In fact, it is known that e is transcendental (that is, it is not a root of any real polynomial with integer coefficients, unlike $\sqrt{2}$, for example). Numerical considerations give $e = 2.71828\ldots$.

5.4 Properties of $\sin z$ and $\cos z$

We turn now to basic properties of the complex trigonometric functions, sin and cos. These must be established from their power series definitions as given above.

For real α, β, we have that $\exp i(\alpha + \beta) = \cos(\alpha + \beta) + i \sin(\alpha + \beta)$. However,

$$\exp i(\alpha + \beta) = \exp i\alpha \, \exp i\beta$$
$$= (\cos\alpha + i\sin\alpha)(\cos\beta + i\sin\beta)$$
$$= \cos\alpha\cos\beta - \sin\alpha\sin\beta + i(\sin\alpha\cos\beta + \cos\alpha\sin\beta).$$

Equating real and imaginary parts leads to the addition rules

$$\sin(\alpha+\beta)=\sin\alpha\cos\beta+\cos\alpha\sin\beta$$

and

$$\cos(\alpha+\beta)=\cos\alpha\sin\beta-\sin\alpha\sin\beta,$$

for any $\alpha,\beta\in\mathbb{R}$. It is natural to ask whether these relationships are also valid for complex α and β. In fact they are, as we now show.

5.5 Addition Formulae

Theorem 5.1 (Addition Formulae) *For any $a,b\in\mathbb{C}$,*

$$\sin(a+b)=\sin a\cos b+\cos a\sin b$$
$$\cos(a+b)=\cos a\cos b-\sin a\sin b.$$

Proof. The easiest way of proving these relations is to write the right hand sides in terms of the exponential function and do a bit of algebra. We find

$$\sin a\cos b+\cos a\sin b$$
$$=\frac{(e^{ia}-e^{-ia})}{2i}\frac{(e^{ib}+e^{-ib})}{2}$$
$$+\frac{(e^{ia}+e^{-ia})}{2}\frac{(e^{ib}-e^{-ib})}{2i}$$
$$=\frac{(e^{ia}e^{ib}+e^{ia}e^{-ib}-e^{-ia}e^{ib}-e^{-ia}e^{-ib})}{4i}$$
$$+\frac{(e^{ia}e^{ib}-e^{ia}e^{-ib}+e^{-ia}e^{ib}-e^{-ia}e^{-ib})}{4i}$$
$$=\frac{(e^{i(a+b)}+e^{i(a-b)}-e^{-i(a-b)}-e^{-i(a+b)})}{4i}$$
$$+\frac{(e^{i(a+b)}-e^{i(a-b)}+e^{-i(a-b)}-e^{-i(a+b)})}{4i}$$
$$=\frac{(e^{i(a+b)}-e^{-i(a+b)})}{2i}$$
$$=\sin(a+b).$$

Similarly, we calculate

$$
\begin{aligned}
\cos a \cos b + \sin a \sin b &= \frac{(e^{ia} + e^{-ia})}{2} \frac{(e^{ib} + e^{-ib})}{2} \\
&\quad + \frac{(e^{ia} - e^{-ia})}{2i} \frac{(e^{ib} - e^{-ib})}{2i} \\
&= \frac{(e^{i(a+b)} + e^{-i(a+b)})}{2} \\
&= \cos(a + b),
\end{aligned}
$$

as required. $\square$

Corollary 5.2 *For any $z \in \mathbb{C}$, $(\sin z)^2 + (\cos z)^2 = 1$.*

Proof. The identity $\cos(a + b) = \cos a \cos b - \sin a \sin b$, with $a = z$ and $b = -z$, gives

$$
\begin{aligned}
\cos 0 &= \cos z \cos(-z) - \sin z \sin(-z) \\
&= \cos z \cos z + \sin z \sin z,
\end{aligned}
$$

where we have used $\cos(-z) = \cos z$ and $\sin(-z) = -\sin z$. The result now follows since $\cos 0 = 1$. $\square$

Remark 5.4 An alternative proof is to set $\psi(z) = \sin^2 z + \cos^2 z$ and then to calculate $\psi'(z)$. One finds that $\psi'(z) = 0$, for all $z \in \mathbb{C}$, and so it follows that ψ is constant. Therefore $\psi(z) = \psi(0) = 1$, as required.

In the same spirit, we can prove the addition formulae by considering the functions

$$
f(z) = \sin(w - z) \cos z + \cos(w - z) \sin z
$$

and

$$
g(z) = \cos(w - z) \cos z - \sin(w - z) \sin z .
$$

One finds that $f'(z) = 0$ and $g'(z) = 0$ for all z, so that both f and g are constant: $f(z) = f(0) = \sin w$ and $g(z) = g(0) = \cos w$. Letting $w = a + b$ and setting $z = b$, we get the required formulae.

Corollary 5.3 *For any $z \in \mathbb{C}$, we have*

$$
\sin 2z = 2 \sin z \cos z \quad \text{and}
$$
$$
\cos 2z = \cos^2 z - \sin^2 z = 2 \cos^2 z - 1 = 1 - 2 \sin^2 z.
$$

Proof. We simply put $a = b = z$ in the addition formulae above and use the identity $\sin^2 z + \cos^2 z = 1$. $\square$

Corollary 5.4   *For any $t \in \mathbb{R}$, $|\exp(it)| = 1$.*

Proof.   We have $\exp(it) = \cos t + i \sin t$. Both $\cos t$ and $\sin t$ are real if t is, and therefore $|\exp(it)|^2 = \cos^2 t + \sin^2 t = 1$. $\square$

Remark 5.5   For real x, $\sin x$ and $\cos x$ are also real, and so the relation $\sin^2 x + \cos^2 x = 1$ shows that $|\sin x| \leq 1$ and also $|\cos x| \leq 1$. These inequalities do not extend to the complex case. For example, if $z = it$, with $t \in \mathbb{R}$, then $\sin z = \sin it = (\exp(-t) - \exp t)/2i$. If t is large (and positive), then $\exp t$ is large, $\exp(-t)$ is small and so $|\sin it|$ is large. A similar remark applies to $\cos z$.

5.6   The Appearance of π

We wish to discuss various properties of $\sin x$ and $\cos x$ for real x. In particular, we would like to introduce the number π.

Theorem 5.2   *If $x \in (0, 2)$, then $\sin x > 0$.*

Proof.   Suppose that $0 < x \leq 1$. Then

$$\cos x = 1 - \frac{x^2}{2!} + \frac{x^4}{4!} - \frac{x^6}{6!} + \dots$$

$$> 1 - \frac{x^2}{2!} - \frac{x^4}{4!} - \frac{x^6}{6!} - \dots$$

$$> 1 - \frac{1}{2} - \frac{1}{2^3} - \frac{1}{2^5} - \dots,$$

since $\dfrac{x^2}{2!} \leq \dfrac{1}{2}$, $\dfrac{x^4}{4!} \leq \dfrac{1}{4.3.2} < \dfrac{1}{2^3}, \dots$ etc.,

$$= 1 - \frac{\frac{1}{2}}{(1 - \frac{1}{4})} = \frac{1}{3}.$$

Hence $\sin' x = \cos x > \frac{1}{3}$ on $[0, 1]$ and so (by the Mean Value Theorem), $\sin x$ is strictly increasing on $[0, 1]$. But $\sin 0 = 0$ and therefore $\sin x > 0$ for $0 < x \leq 1$. It follows that

$$\sin x = 2 \sin \tfrac{x}{2} \cos \tfrac{x}{2} > 0$$

whenever $0 < x \leq 2$, since, in this case, $\sin \frac{x}{2} > 0$ and $\cos \frac{x}{2} > \frac{1}{3}$. $\square$

Corollary 5.5 *The function* $\cos x$ *is strictly decreasing on* $[0,2]$.

Proof. We have $\cos' x = -\sin x$ which is strictly negative on $(0,2)$, by the theorem. By the Mean Value Theorem, it follows that the function $\cos x$ is strictly decreasing on $[0,2]$. $\qquad\square$

Theorem 5.3 $\sin 4 < 0$.

Proof. From the definition of $\sin x$,

$$\sin 4 = 4 - \frac{4^3}{3!} + \frac{4^5}{5!} - \frac{4^7}{7!} + \frac{4^9}{9!}$$
$$- \frac{4^{11}}{11!}\left(1 - \frac{4^2}{12.13}\right) - \frac{4^{15}}{15!}\left(1 - \frac{4^2}{16.17}\right)$$
$$- \frac{4^{19}}{19!}\left(1 - \frac{4^2}{20.21}\right) - \cdots$$
$$< 4 - \frac{4^3}{3!} + \frac{4^5}{5!} - \frac{4^7}{7!} + \frac{4^9}{9!} = -\frac{268}{405}$$

and the result follows. $\qquad\square$

Theorem 5.4 *There is a unique real number* π *satisfying* $0 < \pi < 4$ *such that* $\sin\pi = 0$. *Furthermore,* $\cos\frac{\pi}{2} = 0$, $\sin\frac{\pi}{2} = 1$ *and* $\cos\pi = -1$.

Proof. We have seen that $\sin x > 0$ for $x \in (0,2)$ and so, in particular, $\sin 1 > 0$. We have also shown that $\sin 4 < 0$. Now, the map $x \mapsto \sin x$ is continuous on $\mathbb{R}$, and so, by the Intermediate Value Theorem, there is some real number, which we will denote by π, with $1 < \pi < 4$ and such that $\sin\pi = 0$.

We must now show that π is the only value in $(0,4)$ obeying $\sin\pi = 0$. To this end, suppose that $\sin\alpha = 0$ with $0 < \alpha < 4$. Then

$$0 = \sin\alpha = 2\sin\tfrac{\alpha}{2}\cos\tfrac{\alpha}{2}.$$

Hence either $\sin\frac{\alpha}{2} = 0$ or $\cos\frac{\alpha}{2} = 0$ (they cannot both vanish because the sum of their squares is equal to 1). But if $0 < \alpha < 4$, then $0 < \frac{\alpha}{2} < 2$ and we know that $\sin x > 0$ on $(0,2)$, so $\sin\frac{\alpha}{2}$ cannot be zero. Hence, we must have that $\cos\frac{\alpha}{2} = 0$. In particular, $\cos\frac{\pi}{2} = 0$. However, $\cos x$ is strictly decreasing on $[0,2]$ and so there can be at most one solution to $\cos x = 0$ in this interval. Hence $\frac{\alpha}{2} = \frac{\pi}{2}$ so that $\alpha = \pi$, and the uniqueness is established.

Next, $\sin^2 z + \cos^2 z = 1$ implies that $\sin^2\frac{\pi}{2} = 1$, since $\cos\frac{\pi}{2} = 0$. Hence $\sin\frac{\pi}{2} = 1$ because $\sin\frac{\pi}{2} > 0$ (since $0 < \frac{\pi}{2} < 2$).
Finally, we have $\cos\pi = 2\cos^2\frac{\pi}{2} - 1 = -1$. $\qquad\square$

What's going on?   We have defined the trigonometric functions, out of the blue as it were, by means of power series. This approach avoids any appeal to geometry and right-angled triangles. However, having chosen this route, we must stick with it. In particular, it is necessary to get to π via these definitions rather than by drawing triangles or circles. As we have seen, this is all perfectly possible. Our view here is that the trigonometric functions are as we have defined them. Any results must be deduced as consequences of these power series definitions.

Remark 5.6   Numerical investigation yields $\pi = 3.14159\ldots$. It is known that π is irrational (in fact, transcendental). It is something of a sport (involving some fascinating numerical analysis) to calculate the value of π to a large number of decimal places and this has been done to over a million decimal places. Such programs have been used to test the computational integrity of supercomputers by checking to see whether they get these digits correct or not.

Theorem 5.5   *For any $z \in \mathbb{C}$ and $n \in \mathbb{Z}$,*

$$\sin(z + n\pi) = (-1)^n \sin z$$
$$\cos(z + n\pi) = (-1)^n \cos z.$$

Proof.   We use the trigonometric formulae;

$$\sin(z + \pi) = \sin z \cos \pi + \cos z \sin \pi$$
$$= -\sin z + 0,$$

and

$$\cos(z + \pi) = \cos z \cos \pi - \sin z \sin \pi$$
$$= -\cos z - 0.$$

For $n > 0$, the result now follows by induction. Substituting $z = w - n\pi$, the result then follows for $n < 0$. $\square$

Remark 5.7   Particular cases of the above formulae deserve mention. If we set $z = 0$, then we see that $\sin(n\pi) = 0$ and $\cos(n\pi) = (-1)^n$, for any $n \in \mathbb{Z}$. The functions sin and cos are periodic (with period 2π). In particular, for any $x \in \mathbb{R}$ and any $n \in \mathbb{Z}$,

$$\sin(x + 2n\pi) = \sin x \quad \text{and}$$
$$\cos(x + 2n\pi) = \cos x.$$

If we piece together all the information gained above, we recover the familiar picture of $\sin x$ and $\cos x$, for $x \in \mathbb{R}$, as periodic 'wavy' functions. In fact, $\sin x$ is an odd function, so it is determined by its values on $x \geq 0$. It is periodic, so it is determined by its values on $[0, 2\pi]$. But, $\sin(x + \pi) = \sin x \cos \pi + \cos x \sin \pi = -\sin x$ and so $\sin x$ is determined by its values on the interval $[0, \pi]$. Now,

$$\sin\left(\tfrac{\pi}{2} \pm x\right) = \sin \tfrac{\pi}{2} \cos x \pm \cos \tfrac{\pi}{2} \sin x = \cos x,$$

so we see that $\sin x$ is symmetric about $x = \tfrac{\pi}{2}$. It follows that $\sin x$ is completely determined by its values on $[0, \tfrac{\pi}{2}]$.

Furthermore, $\sin\left(x + \tfrac{\pi}{2}\right) = \cos x$ and so the graph of $\cos x$ is got by translating the graph of $\sin x$ by $\tfrac{\pi}{2}$ to the left.

5.7   Inverse Trigonometric Functions

From the analysis above, we see that $\sin x$ is strictly increasing on the interval $[-\tfrac{\pi}{2}, \tfrac{\pi}{2}]$, $\cos x$ is strictly decreasing on the interval $[0, \pi]$ and is strictly increasing on the interval $[-\pi, 0]$. In particular, this means that $\sin$ is a one-one map of $[-\tfrac{\pi}{2}, \tfrac{\pi}{2}]$ onto $[-1, 1]$, $\cos$ is a one-one map of $[0, \pi]$ onto $[-1, 1]$ and also of $[-\pi, 0]$ onto $[-1, 1]$. For $t \in [-1, 1]$, let $\phi(t)$ be the unique element of $[-\tfrac{\pi}{2}, \tfrac{\pi}{2}]$ such that $\sin \phi(t) = t$, let $\psi(t)$ be the unique element of $[0, \pi]$ such that $\cos \psi(t) = t$ and let $\rho(t)$ be the unique element of $[-\pi, 0]$ with $\cos \rho(t) = t$. Thus, ϕ is the inverse of $\sin : [-\tfrac{\pi}{2}, \tfrac{\pi}{2}] \to [-1, 1]$, ψ is the inverse of $\cos : [0, \pi] \to [-1, 1]$ and ρ is the inverse of $\cos : [-\pi, 0] \to [-1, 1]$.

The standard inverse trigonometric functions $\sin^{-1}$ and $\cos^{-1}$ are given by $\sin^{-1}(t) = \phi(t)$ and $\cos^{-1}(t) = \psi(t)$ for $t \in [-1, 1]$ so that $\sin^{-1}$ takes values in $[-\tfrac{\pi}{2}, \tfrac{\pi}{2}]$ whilst $\cos^{-1}$ takes values in $[0, \pi]$.

Theorem 5.6   *Suppose that $f : [a, b] \to [c, d]$ is a strictly increasing (or decreasing) continuous map from $[a, b]$ onto $[c, d]$. Then the inverse map $f^{-1} : [c, d] \to [a, b]$ is continuous.*

Proof.   The idea of the proof is straightforward, but it is a nuisance having to consider the end-points c and d of $[c, d]$ and the interior (c, d) separately. To avoid this, we shall first consider the case of $f : \mathbb{R} \to \mathbb{R}$. Suppose, then, that $f : \mathbb{R} \to \mathbb{R}$ is, say, strictly increasing and maps $\mathbb{R}$ onto $\mathbb{R}$. Let $y_0 \in \mathbb{R}$ and $\varepsilon > 0$ be given. Let $x_0 \in \mathbb{R}$ be the unique point such that $f(x_0) = y_0$, thus, $x_0 = f^{-1}(y_0)$. Set $y_1 = f(x_0 - \varepsilon)$ and $y_2 = f(x_0 + \varepsilon)$,

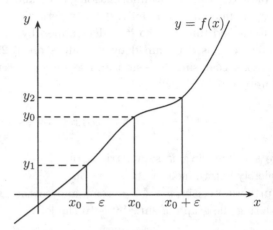

Fig. 5.1   Continuity of the inverse function.

as shown in Fig. 5.1. Then $y_1 < y_0 < y_2$. Furthermore, if $y_1 < y < y_2$, then $x_0 - \varepsilon < f^{-1}(y) < x_0 + \varepsilon$. Putting $\delta = \min\{y_0 - y_1, y_2 - y_0\}$ we see that $|y - y_0| < \delta$ implies that $\left|f^{-1}(y) - f^{-1}(y_0)\right| < \varepsilon$. That is, f^{-1} is continuous at y_0 and therefore on all of $\mathbb{R}$.

The case where f is strictly decreasing is similarly proved. Alternatively, one can note that $f^{-1}(y) = (-f)^{-1}(-y)$ and that $-f$ is increasing if f is decreasing.

Returning now to the case where f is strictly increasing and maps $[a, b]$ onto $[c, d]$, we simply extend f to the whole of $\mathbb{R}$. Define $F : \mathbb{R} \to \mathbb{R}$ by

$$F(x) = \begin{cases} x + f(a) - a, & x < a \\ f(x), & a \leq x \leq b \\ x + f(b) - b, & b < x. \end{cases}$$

Then F maps $\mathbb{R}$ onto $\mathbb{R}$, is strictly increasing, and is equal to f on $[a, b]$. As above, we see that F^{-1} is continuous. In particular, $F^{-1} = f^{-1}$ is continuous on $[c, d]$.   □

Theorem 5.7   *Each of the functions ϕ, ψ, and ρ is continuous.*

Proof.   This follows immediately from the preceding theorem.   □

5.8 More on $\exp z$ and the Zeros of $\sin z$ and $\cos z$

Having established the familiar properties of the real trigonometric functions, we return to a discussion of the complex versions. We attack these via the exponential function, which determines them all.

Proposition 5.3 *The equality* $\exp z = \exp w$, *for* $z, w \in \mathbb{C}$, *holds if and only if there is some* $k \in \mathbb{Z}$ *such that* $z = w + 2\pi k i$.

Proof. Multiplying both sides by $\exp(-w)$, we see that $\exp z = \exp w$ if and only if $\exp(z - w) = 1$. Now, $\exp(2\pi k i) = \cos 2\pi k + i \sin 2\pi k = 1$ and therefore $\exp z = \exp w$ if $z - w = 2\pi k i$.

Conversely, suppose that $\exp(z - w) = 1$, and write $z - w = \alpha + i\beta$, with $\alpha, \beta \in \mathbb{R}$. Then

$$1 = \exp(z - w) = \exp(\alpha + i\beta)$$
$$= \exp \alpha \, \exp i\beta$$
$$= \exp \alpha \, (\cos \beta + i \sin \beta).$$

It follows that $\exp \alpha = 1$ (taking the modulus of both sides), and so $\cos \beta = 1$ and $\sin \beta = 0$ (equating real and imaginary parts). This implies that $\alpha = 0$ and β is of the form $\beta = 2\pi k$, for some $k \in \mathbb{Z}$. It follows that $z - w = 2\pi k i$, for some $k \in \mathbb{Z}$, as required. $\qquad\square$

We have extended the definition of the functions $\sin x$ and $\cos x$ from the real variable x to the complex variable z. It is of interest to note that these complex trigonometric functions have no new zeros, as we show next.

Proposition 5.4 *For* $z \in \mathbb{C}$, $\sin z = 0$ *if and only if* $z = \pi k$ *for some* $k \in \mathbb{Z}$, *and* $\cos z = 0$ *if and only if* $z = (2k + 1)\frac{\pi}{2}$, *for some* $k \in \mathbb{Z}$.

Proof. We have

$$\sin z = 0$$
$$\iff \frac{\exp(iz) - \exp(-iz)}{2i} = 0$$
$$\iff \exp(iz) = \exp(-iz)$$
$$\iff \exp(2iz) = \exp 0$$
$$\iff 2iz = 0 + 2\pi k i, \text{ for some } k \in \mathbb{Z},$$
$$\iff z = \pi k, \text{ for some } k \in \mathbb{Z},$$

as claimed.

A similar argument is used for $\cos z$. Indeed,

$$\cos z = 0$$

$$\Longleftrightarrow \frac{\exp(iz) + \exp(-iz)}{2} = 0$$

$$\Longleftrightarrow \exp(iz) = -\exp(-iz)$$

$$\Longleftrightarrow \exp(2iz) = -1$$

$$\Longleftrightarrow \exp(2iz) = \exp(i\pi), \text{ since } \exp(i\pi) = -1,$$

$$\Longleftrightarrow 2iz = i\pi + 2\pi ki, \text{ for some } k \in \mathbb{Z},$$

$$\Longleftrightarrow z = (2k+1)\tfrac{\pi}{2}, \text{ for some } k \in \mathbb{Z},$$

and the proof is complete.   □

5.9   The Argument Revisited

We know, informally, that any non-zero complex number can be written as $r(\cos\theta + i\sin\theta) = re^{i\theta}$, where r is its modulus and θ is some choice of argument (angle with the positive real axis). We shall establish this formally and also consider the argument mapping in more detail.

Suppose, then, that $z \neq 0$. Write $z = a + ib$, with $a, b \in \mathbb{R}$. Then $r = |z| = \sqrt{a^2 + b^2} \neq 0$ and $z = r(\alpha + i\beta)$ where $\alpha = a/r$, $\beta = b/r$. Clearly, $\alpha^2 + \beta^2 = 1$ and so $|\alpha| \leq 1$ and $|\beta| \leq 1$. We would like to show that it is possible to find θ such that $\alpha = \cos\theta$ and $\beta = \sin\theta$. Since $|\beta| \leq 1$, there is a unique $\theta \in [-\tfrac{\pi}{2}, \tfrac{\pi}{2}]$ such that $\sin\theta = \beta$. Now, $\cos^2\theta = 1 - \sin^2\theta = 1 - \beta^2 = \alpha^2$. It follows that $\cos\theta = \pm\alpha$. If $\cos\theta = \alpha$, we have $z = re^{i\theta}$ and we are done. (Note that $\cos\theta \geq 0$ since $-\tfrac{\pi}{2} \leq \theta \leq \tfrac{\pi}{2}$.) If not, we must have $\cos\theta = -\alpha$. Let $\theta' = \pi - \theta$. Then, using the trigonometric formulae above, we find that $\sin\theta' = \sin\pi\cos\theta - \cos\pi\sin\theta = \sin\theta$ and $\cos\theta' = \cos\pi\cos\theta + \sin\pi\sin\theta = -\cos\theta = \alpha$. Therefore, in this case, we can write z as $z = re^{i\theta'}$.

Consider the family of complex numbers given by $z(t) = e^{2\pi it} = \cos 2\pi t + i\sin 2\pi t$, for $t \geq 0$. Clearly, $|z(t)| = 1$ and $z(0) = 1$. As we continuously increase t, the complex number $z(t)$ moves continuously anti-clockwise around the circle centred on the origin and with radius one. For $t = \tfrac{1}{8}$, we find $z(\tfrac{1}{8}) = (1+i)/\sqrt{2}$. Also, we have $z(\tfrac{1}{4}) = i$, $z(\tfrac{1}{2}) = -1$, $z(\tfrac{3}{4}) = -i$ and $z(1) = 1$.

We have seen that $\exp z = \exp w$ if and only if there is some integer $k \in \mathbb{Z}$ such that $w = z + 2\pi ki$. It follows that we can always write $z = re^{i\theta}$

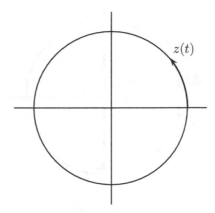

Fig. 5.2 The point $z(t) = e^{2\pi it}$ moves anticlockwise around the unit circle.

where the argument $\theta \in \mathbb{R}$ is only defined up to additional integer multiples of 2π, i.e., $z = re^{i\theta} = re^{i\lambda}$ if and only if $\lambda = \theta + 2\pi k$ for some $k \in \mathbb{Z}$. In particular, we can always find a unique value for θ in the range $(-\pi, \pi]$. (The difference between successive possible values for θ is 2π, so there must be exactly one such value in any open-closed interval of length 2π.) This value of the argument is given a special name, as already discussed (though somewhat informally).

Definition 5.3 For any $z \neq 0$, the principal value of the argument of z is the unique real number $\operatorname{Arg} z$ satisfying $-\pi < \operatorname{Arg} z \leq \pi$ and such that $z = |z| \, e^{i \operatorname{Arg} z}$. ($\operatorname{Arg} z$ is not defined for $z = 0$.)

Notice that if z is close to the negative real axis, then its imaginary part is small and its real part is negative. If its imaginary part is positive, then the principal value of its argument is close to π, whereas if its imaginary part is negative then the principal value of its argument is close to $-\pi$.

In the example above, the principal value of the argument of the complex number $z(t)$ increases from 0, when $t = 0$, through $\frac{\pi}{2}$, when $t = \frac{1}{4}$, to π, when $t = \frac{1}{2}$ and $z = -1$. However, as $z(t)$ crosses the negative real axis, from above to below, the principal value of its argument jumps from π to "nearly" $-\pi$ (it never assumes the value $-\pi$). (The limit from above is π, whereas the limit from below is $-\pi$.) It continues to increase, as t increases, until it has the value 0, when $t = 1$ and $z = 1$. The negative real axis is a line of discontinuity for $\operatorname{Arg} z$. (Recall that $\operatorname{Arg} z$ is not defined if $z = 0$.)

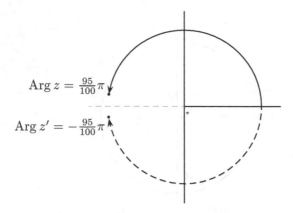

$$\text{Arg } z = \tfrac{95}{100}\pi$$

$$\text{Arg } z' = -\tfrac{95}{100}\pi$$

Fig. 5.3   Arg z is discontinuous across the negative real axis.

5.10   Arg z is Continuous in the Cut-Plane

We shall show next that Arg z is continuous everywhere apart from on the negative real-axis.

Theorem 5.8   *The map $z \mapsto \text{Arg } z$ is a continuous mapping from the cut-plane $\mathbb{C} \setminus \{\, z : z + |z| = 0 \,\}$ onto the interval $(-\pi, \pi)$.*

Proof.   Let $z \in \mathbb{C} \setminus \{\, z : z + |z| = 0 \,\}$, i.e., z is any point of $\mathbb{C}$ not on the negative real axis (including 0). Write $z = x + iy = r(\cos \text{Arg } z + i \sin \text{Arg } z)$, where $r = |z|$. Consider first the region where $\text{Im } z = y = r \sin \text{Arg } z > 0$, that is, the upper half-plane. Then $\sin \text{Arg } z > 0$ and so $0 < \text{Arg } z < \pi$. But $r \cos \text{Arg } z = x$ and so $\text{Arg } z = \psi(x/r)$ where ψ is the inverse to $\cos : [0, \pi] \to [-1, 1]$. But this map is continuous, by theorem 5.7, and so $\text{Arg } z$ is continuous on the upper half-plane $\{\, z : \text{Im } z > 0 \,\}$

Next consider the region $x = \text{Re } z > 0$ (right half-plane). For such z, we have $\cos \text{Arg } z > 0$, so that $-\tfrac{\pi}{2} < \text{Arg } z < \tfrac{\pi}{2}$. However, $r \sin \text{Arg } z = y$ and therefore $\text{Arg } z = \phi(y/r)$, where ϕ is the inverse to $\sin : [-\tfrac{\pi}{2}, \tfrac{\pi}{2}] \to [-1, 1]$. Again, by theorem 5.7, we conclude that $\text{Arg } z$ is continuous on this region, namely the right half-plane $\{\, z : \text{Re } z > 0 \,\}$.

Finally, consider the region with $y = \text{Im } z < 0$. Here, $\sin \text{Arg } z < 0$, so that $-\pi < \text{Arg } z < 0$. Since $r \cos \text{Arg } z = x$, it follows that $\text{Arg } z = \rho(x/r)$, where ρ is the inverse to $\cos : [-\pi, 0] \to [-1, 1]$. Again, by theorem 5.7, we see that $\text{Arg } z$ is continuous on $\{\, z : \text{Im } z < 0 \,\}$, the lower half-plane. We conclude that $z \mapsto \text{Arg } z$ is continuous on $\mathbb{C} \setminus \{\, z : z + |z| = 0 \,\}$.

If $z \neq 0$ does not lie on the negative real-axis, then $\operatorname{Arg} z \neq \pi$ and so $\operatorname{Arg} z \in (-\pi, \pi)$. On the other hand, for any $-\pi < \theta < \pi$, we have $\operatorname{Arg}(\cos\theta + i\sin\theta) = \theta$. It follows that $z \mapsto \operatorname{Arg} z$ maps $\mathbb{C} \setminus \{\, z : z + |z| = 0 \,\}$ onto the interval $(-\pi, \pi)$. $\qquad\qquad\qquad\qquad\qquad\qquad\qquad\qquad\square$

What's going on?   The claimed continuity of the function $\operatorname{Arg} z$ on the cut-plane $\mathbb{C} \setminus \{\, z : z + |z| = 0 \,\}$ is clear from a diagram. However, our starting point has been the power series definitions of the trigonometric functions and the function $\operatorname{Arg} z$ is formally constructed in terms of these (via suitable inverses). This has meant that we have had to do some work to wring out the required (but "obvious") behaviour.

Chapter 6

The Complex Logarithm

6.1 Introduction

The logarithm is an inverse for the exponential function: if $x = e^t$, then $\ln x = t$, where $t \in \mathbb{R}$. Note that the natural logarithm $\ln x$ is very often also written as $\log x$. Can we mimic this construction to get a logarithm for complex variables?

If $z \in \mathbb{C}$, we want $\log z = w$, whatever it turns out to be, to satisfy the relation $e^w = z$. To see how we might proceed, write z as $z = x + iy = re^{i\theta}$, with $r = \sqrt{x^2 + y^2}$ and note that θ is not uniquely determined—we can always add $2\pi k$, for any $k \in \mathbb{Z}$. Nevertheless, suppose that we have made a choice for θ. Then we want to construct $\log z = \log(re^{i\theta})$.

We try the formula

$$\log z = \log(r\,e^{i\theta}) = \underbrace{\log r}_{\text{usual log}} + \underbrace{\log e^{i\theta}}_{\text{undefined as yet}}$$

$$= \ln r + i\theta$$

as an apparently reasonable attempt. Then we find that whatever our choice for θ

$$e^{\ln r + i\theta} = e^{\ln r}\,e^{i\theta} = r\,e^{i\theta} = z,$$

since $r > 0$ and so $e^{\ln r} = r$. In fact, if $w = \ln|z| + i\theta + i2\pi k$ then

$$e^w = e^{\ln|z| + i\theta + i2\pi k}$$

$$= e^{\ln|z|}\,e^{i\theta}\,e^{i2\pi k}$$

$$= |z|\,e^{i\theta}e^{i2\pi k} = ze^{i2\pi k}$$

$$= z,$$

for any $k \in \mathbb{Z}$. In other words, for given $z \neq 0$, the equation

$$e^w = z$$

has infinitely-many solutions, namely,

$$w = \ln|z| + i\arg z,$$

where $\arg z$ is any real number such that $z = |z|\, e^{i\arg z}$. For $z = 0$, the equation becomes $e^w = 0$, which has no solution. (We know that the complex exponential function is never zero.) These heuristics suggest that setting up a theory of complex logarithms might be a little more involved than for the real case.

6.2   The Complex Logarithm and its Properties

Definition 6.1   For $z \in \mathbb{C} \setminus \{0\}$, we say that a logarithm of z is any particular solution w to $e^w = z$.

If w_1 and w_2 are solutions to $e^w = z$, then $e^{w_1} = z = e^{w_2}$ and so we see that $e^{w_1 - w_2} = 1$. Putting $w_1 - w_2 = a + ib$, with $a, b \in \mathbb{R}$, this becomes $e^a e^{ib} = 1$ and so $a = 0$ and $b = 2\pi k$, for some $k \in \mathbb{Z}$. In other words, the difference between any two possible choices for the logarithm is always an integer multiple of $2\pi i$.

The arbitrariness of the complex logarithm reflects the ambiguity in the choice of the argument of a complex number. When there is no chance of confusion, one often just writes $\log r$ to mean the usual real logarithm of any positive real number r.

Definition 6.2   The principal value of the logarithm of $z \neq 0$ is that obtained via the principal value of the argument and is denoted $\operatorname{Log} z$; thus

$$\operatorname{Log} z = \ln|z| + i\operatorname{Arg} z.$$

We see that $-\pi < \operatorname{Im} \operatorname{Log} z \leq \pi$. Moreover, since possible choices of the argument must differ by some integer multiple of 2π, it follows that *any* choice of logarithm of z, $\log z$, can be written as $\log z = \operatorname{Log} z + i2\pi k$, for some $k \in \mathbb{Z}$. Of course, k may depend on z.

Various properties of any such choice, $\log z$, of the logarithm of z are considered next.

Theorem 6.1 *Suppose that for each $z \in \mathbb{C} \setminus \{0\}$ a value for $\log z$ has been chosen. Then the following hold.*

(i) $e^{\log z} = z.$

(ii) $\log(e^z) = z + 2\pi ki$, *for some* $k \in \mathbb{Z}.$

(iii) $\log(z_1 z_2) = \log z_1 + \log z_2 + 2\pi ki$, *for some* $k \in \mathbb{Z}.$

(iv) $\log\left(\dfrac{1}{z}\right) = -\log z + 2\pi ki$, *for some* $k \in \mathbb{Z}.$

Proof. By definition, any choice $w = \log z$ of the logarithm of z satisfies $e^w = z$, which is (i).

To prove (ii), let $w = \log(e^z)$ be any choice of the logarithm of e^z. Then $e^w = e^z$. It follows that $w - z = 2\pi ki$, for some $k \in \mathbb{Z}$, as required.

Let $w_1 = \log z_1$ and $w_2 = \log z_2$ be any choices of the logarithms of z_1 and z_2, respectively, and let $w_3 = \log(z_1 z_2)$ be some choice for the logarithm of the product $z_1 z_2$. By the definition,

$$e^{w_3} = z_1 z_2 = e^{w_1} e^{w_2} = e^{w_1 + w_2}.$$

It follows that there is some $k \in \mathbb{Z}$ such that $w_3 = w_1 + w_2 + 2\pi ki$, which proves part (iii).

Finally, suppose that w is a choice of $\log \frac{1}{z}$ and let ζ be any choice of $\log z$. Then $e^w = 1/z$ and $e^\zeta = z$. It follows that $z = 1/e^w$ and therefore

$$e^\zeta = z = \frac{1}{e^w} = e^{-w}.$$

We deduce that $\zeta = -w + 2\pi ki$, for suitable $k \in \mathbb{Z}$, which proves (iv). $\square$

Examples 6.1

(1) Possible choices of the logarithm of 1 are $\log 1 = 0$, or $2\pi i$, or $4\pi i$, or $\ldots$, or $-2\pi i$, or $-4\pi i$, or $\ldots$.

(2) $\log 1 = \log 1^2 \overset{?}{=} \log 1 + \log 1$. This is only true if we make the choice $\log 1 = 0$. No choice $\log 1 = 2\pi ki$, with $k \in \mathbb{Z}$, $k \neq 0$, will work.

(3) Consider the principal value of the logarithm of the product $i(-1 + i)$. We have

$$\operatorname{Log} i(-1 + i) = \ln |i(-1 + i)| + i \operatorname{Arg} i(-1 + i)$$
$$= \ln \sqrt{2} - i \tfrac{3\pi}{4}.$$

Now, $\operatorname{Log} i = \ln |i| + i \operatorname{Arg} i = 0 + i\frac{\pi}{2}$ and $\operatorname{Log}(-1+i)$ is given by

$$\operatorname{Log}(-1+i) = \ln |-1+i| + i \operatorname{Arg}(-1+i)$$
$$= \ln \sqrt{2} + i \frac{3\pi}{4}.$$

We see that

$$\operatorname{Log} i + \operatorname{Log}(-1+i) = i\frac{\pi}{2} + \ln \sqrt{2} + i\frac{3\pi}{4} = \ln \sqrt{2} + i\frac{5\pi}{4}$$
$$\neq \operatorname{Log} i(-1+i).$$

(4) For which values of z is $\sin z = 3$? To solve this, put $w = e^{iz}$, so that $\sin z = (w - w^{-1})/2i$. Then $\sin z = 3$ becomes the quadratic equation $w^2 - 6iw - 1 = 0$ with solutions $w = (3 \pm 2\sqrt{2})i$. But $e^{iz} = w$ means that iz is a choice of $\log w$, that is, iz must be of the form $iz = \operatorname{Log} w + 2k\pi i$ for some $k \in \mathbb{Z}$. Hence $iz = \ln |3 \pm 2\sqrt{2}| + i\frac{\pi}{2} + 2k\pi i$ and so $z = -i \ln |3 \pm 2\sqrt{2}| + \frac{\pi}{2} + 2k\pi$ for $k \in \mathbb{Z}$.

Remark 6.1   If z is real and positive, $z = x + iy$, $x > 0$ and $y = 0$, then

$$\operatorname{Log} x = \operatorname{Log} z = \ln |z| + i \operatorname{Arg} z = \ln x + i0$$
$$= \ln x.$$

In other words, on the positive real axis, $\{\, z : z \text{ real and strictly positive}\,\}$, the principal value of the logarithm agrees with the usual real logarithm.

6.3   Complex Powers

For real numbers, a and b, with $a > 0$, the definition of the power a^b is given by $a^b = \exp(b \log a)$. Now that we have a meaning of the logarithm for complex numbers we can try to similarly define complex powers.

Definition 6.3   For given $z, \zeta \in \mathbb{C}$, with $z \neq 0$, we define the power

$$z^\zeta = \exp(\zeta \log z).$$

Evidently, z^ζ depends on the choice of the logarithm $\log z$. That is, we must first make a choice of $\log z$ before we can define z^ζ. Put another way, different choices of $\log z$ will lead to different values for z^ζ. The principal value of z^ζ, for $z \neq 0$, is defined to be $\exp(\zeta \operatorname{Log} z)$.

Examples 6.2

(1) What are the possible values of $8^{\frac{1}{3}}$? We have

$$8^{\frac{1}{3}} = e^{\frac{1}{3} \log 8} = e^{\frac{1}{3}(\ln 8 + 2\pi k i)}$$
$$= e^{\ln 2 + \frac{2}{3}\pi k i} = 2e^{\frac{2}{3}\pi k i} ,$$

for $k \in \mathbb{Z}$. Taking $k = 0, 1, 2$ gives all the possibilities (further choices of k merely give repetitions of these three values).

(2) The possible values of i^i are

$$i^i = e^{i \log i} = e^{i (\operatorname{Log} i + 2k\pi i)}, \quad \text{for } k \in \mathbb{Z},$$
$$= e^{i(i\frac{\pi}{2} + 2k\pi i)}$$
$$= e^{-\frac{\pi}{2} - 2k\pi}, \quad \text{for } k \in \mathbb{Z},$$

which are all real.

(3) Taking principal values, $(-i)^{1/2} = e^{\frac{1}{2} \operatorname{Log}(-i)} = e^{-i\pi/4}$, $(-1)^{1/2} = e^{\frac{1}{2} \operatorname{Log}(-1)} = e^{i\pi/2}$ and $i^{1/2} = e^{\frac{1}{2} \operatorname{Log} i} = e^{i\pi/4}$, so that

$$(-i)^{1/2} = e^{-i\pi/4} \neq e^{3i\pi/4} = (-1)^{1/2} i^{1/2}.$$

This provides an example of complex numbers w and ζ for which

$$(w\,\zeta)^{1/2} \neq w^{1/2} \zeta^{1/2}$$

where the principal value of the square root is taken.

Remark 6.2 There are one or two consistency issues to worry about. We seem to have two possible meanings of z^m when $m \in \mathbb{Z}$, namely, as the product of z with itself m times (or the inverse of this if m is negative) or as the quantity $\exp(m \log z)$, for some choice of $\log z$. In fact, there is no need to worry. Let $\log z$ be any fixed choice of the logarithm of z (where $z \neq 0$). Suppose that $m \in \mathbb{N}$. Then

$$\exp(m \log z) = \underbrace{\exp(\log z) \dots \exp(\log z)}_{m \text{ factors}} = \underbrace{z \dots z}_{m \text{ factors}} ,$$

which shows that $\exp(m \log z)$ reduces to the usual "product of m terms" meaning of z^m. For negative m, set $k = -m$. Then, as above,

$$\exp(m \log z) = \exp(-k \log z) = \frac{1}{\exp(k \log z)}$$

$$= \frac{1}{\underbrace{z \times \cdots \times z}_{k \text{ terms}}}$$

$$= z^{-k}, \text{ usual meaning}$$

$$= z^m, \text{ usual meaning}.$$

For $m = 0$, we have $\exp(m \log z) = \exp 0 = 1 = z^0$ which is the usual meaning of a number to the zeroth power.

Another concern is with our agreed notation e^z for $\exp z$. Does this conflict with the meaning of the real number e raised to the complex power of z? From the definition, we see that $e^z = \exp(z \log e)$ with some choice of the logarithm being made. Now, any such choice has the form $\log e = \operatorname{Log} e + 2\pi ki$, for some suitable $k \in \mathbb{Z}$. This means that we can always write e^z as $\exp(z(\operatorname{Log} e + 2\pi ki)) = \exp(z + 2z\pi ki) = \exp z \exp(2z\pi ki)$, since $\operatorname{Log} e = 1$. For this to equal $\exp z$, we must insist that $\exp(2z\pi ki) = 1$, that is, $2z\pi ki = 2\pi mi$ for some $m \in \mathbb{Z}$. In general, this is only possible when $k = 0$, in which case $m = 0$.

We come to the conclusion that the complex power e^z agrees with $\exp z$ provided that we always take the principal value of the power. We shall adopt this convention if there is any doubt. In fact, e^z is really only usually used as a notational shorthand for $\exp z$. So we can choose not to use it, and always use $\exp z$ instead, or to use it and remember exactly what we are doing, or use it and always interpret it as the principal value of the power. There is unlikely ever to be any confusion.

Proposition 6.1 *Suppose that $z \neq 0$ and for $m \in \mathbb{N}$ let $\zeta = z^{1/m}$. Then $\zeta^m = z$, that is, ζ is an mth root of z.*

Proof.   We have that $\zeta = \exp(\frac{1}{m} \log z)$ for some choice of $\log z$. But then

$$\zeta^m = \underbrace{\exp(\frac{1}{m} \log z) \ldots \exp(\frac{1}{m} \log z)}_{m \text{ terms}} = \exp \log z = z,$$

as required.   □

Example 6.3 Let a be any choice of $\sqrt{-1}$ (i.e., of $(-1)^{\frac{1}{2}}$) and let b be any choice of $\sqrt{1}$ (i.e., of $1^{\frac{1}{2}}$). Then

$$\sqrt{\frac{-1}{1}} = \sqrt{\frac{1}{-1}} = \sqrt{-1} = a.$$

But

$$\frac{\sqrt{-1}}{\sqrt{1}} = \frac{a}{b} \text{ and } \frac{\sqrt{1}}{\sqrt{-1}} = \frac{b}{a}.$$

For equality, i.e., for $\sqrt{\frac{-1}{1}} = \frac{\sqrt{-1}}{\sqrt{1}} = \frac{\sqrt{1}}{\sqrt{-1}} = \sqrt{\frac{1}{-1}}$, we require that

$$a = \frac{a}{b} = \frac{b}{a}.$$

This, in turn, requires that $a^2 = b^2$ or $-1 = 1$, which evidently can never hold (no matter what choices a and b are made). We conclude that whilst

$$\sqrt{\frac{-1}{1}} = \sqrt{\frac{1}{-1}},$$

since both are just some choice of $\sqrt{-1}$, nevertheless

$$\frac{\sqrt{-1}}{\sqrt{1}} \neq \frac{\sqrt{1}}{\sqrt{-1}}.$$

This means that the "equalities"

$$\sqrt{\frac{-1}{1}} = \frac{\sqrt{-1}}{\sqrt{1}} \text{ and } \sqrt{\frac{1}{-1}} = \frac{\sqrt{1}}{\sqrt{-1}}$$

simply cannot ever both be simultaneously true.

6.4 Branches of the Logarithm

We have seen that $\log z$ depends on a choice for $\arg z$. A natural question to ask is whether or not there is some consistent choice of $\arg z$ so as to make $z \mapsto \log z$ continuous. The answer depends on where exactly we want to define $\log z$, i.e., its domain of definition. For some regions, for example an annulus around the origin such as $\{ z : 1 < |z| < 2 \}$, this cannot be done, as we will show below.

Proposition 6.2 *The choice $z \mapsto \operatorname{Log} z$ of the logarithm is continuous on the cut plane $\mathbb{C} \setminus \{\, z : z + |z| = 0 \,\}$.*

Proof. For $z \in \mathbb{C} \setminus \{\, z : z + |z| = 0 \,\}$, $\operatorname{Log} z = \ln |z| + i \operatorname{Arg} z$. Now, we have seen that $\operatorname{Arg} z$ is continuous at each such z and so is $z \mapsto |z|$ and hence also $z \mapsto \ln |z|$. The result follows. $\qquad \square$

Example 6.4 There is no choice of logarithm making the map $z \mapsto \log z$ continuous everywhere on the circle $C = \{\, z : |z| = r \,\}$. To see this, suppose that $z \mapsto f(z)$ is such a choice of logarithm, for $z \in C$. We know that $z \mapsto \operatorname{Log} z$ is continuous for z not on the negative real axis, and so the map $z \mapsto f(z) - \operatorname{Log} z$ is continuous for $z \in C \setminus \{-r\}$.

Now, we have $z = e^{f(z)}$, by definition of a logarithm. But $z = e^{\operatorname{Log} z}$ and so $e^{f(z)} = e^{\operatorname{Log} z}$ for $z \in C$. It follows that $f(z) = \operatorname{Log} z + 2\pi i k(z)$ for some $k(z) \in \mathbb{Z}$, depending possibly on $z \in C$. However, both $f(z)$ and $\operatorname{Log} z$ are continuous on $C \setminus \{-r\}$ and so, therefore, is their difference $f(z) - \operatorname{Log} z = 2\pi i k(z)$.

For $-\pi < t < \pi$, set $z(t) = r(\cos t + i \sin t) = re^{it}$. Then the map $t \mapsto k(z(t))$ is a continuous map from $(-\pi, \pi) \to \mathbb{Z}$ and so must be constant (by the Intermediate Value Theorem). We conclude that there is some fixed $k \in \mathbb{Z}$ such that

$$f(z) = \operatorname{Log} z + 2\pi i k = \ln r + i \operatorname{Arg} z + 2\pi i k \qquad (*)$$

for all $z \in C \setminus \{-r\}$. However, the left hand side of $(*)$ is continuous at each $z \in C$, by hypothesis, whereas the right hand side has no limit as z approaches $-r$. (If z approaches $-r$ from above (i.e., through positive imaginary parts) then $\operatorname{Arg} z$ converges to π, but if z approaches $-r$ from below, then $\operatorname{Arg} z$ converges to $-\pi$.) This contradiction shows that such a continuous choice of logarithm on C cannot be made.

Definition 6.4 A branch of the logarithmic function is a pair (D, f), where D is a domain and $f : D \to \mathbb{C}$ is continuous and satisfies $\exp f(z) = z$ for all $z \in D$. (Note that D cannot contain 0 since the exponential function never vanishes.)

The principal branch is that with $D = \mathbb{C} \setminus \{\, z : z + |z| = 0 \,\}$ and $f(z) = \operatorname{Log} z$.

By suitably modifying $\operatorname{Log} z$ in various regions of the complex plane, we can construct other branches of the logarithm.

Example 6.5 Let $D = \mathbb{C} \setminus \{\, z : z - |z| = 0 \,\}$, the complex plane with the positive real axis (and $\{0\}$) removed. We define $f(z)$, for $z \in D$, in terms of Log z, as indicated in Fig. 6.1. For given $z \in D$, set

$$f(z) = \begin{cases} \text{Log } z, & \text{Im } z \geq 0 \\ \text{Log } z + 2\pi i, & \text{Im } z < 0. \end{cases}$$

Notice that $f(z)$ takes the value Log z for z on the negative real axis.

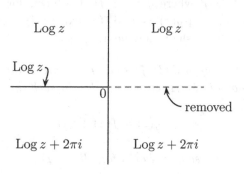

Fig. 6.1   A branch of the logarithm on $\mathbb{C} \setminus \{ z : z - |z| = 0 \}$.

It is clear that f, as defined here, is continuous at any z not on the real axis. The function Log z jumps by $-2\pi i$ on crossing the negative real axis from above to below, so the construction of f, via the addition of an extra $2\pi i$ in the lower half-plane, ensures its continuity, even on the negative real axis.

Example 6.6   Let S be the set $S = \{ z : z = t + it\,(t - 1),\ t \geq 0 \}$ and let $D = \mathbb{C} \setminus S$. By way of example, we seek a branch (D, f) of the

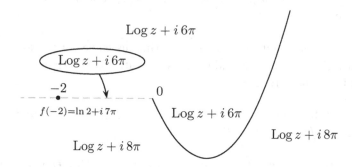

Fig. 6.2   A branch of log z on D with $f(-2) = \ln 2 + i\,7\pi$.

logarithm on the domain such that $f(-2) = \ln 2 + i\,7\pi$. Notice that D is the complement of part of a parabola. (Let $x = t$ and $y = t(t-1)$ so that $y = x(x-1)$ for $x \geq 0$.) The branch (D, f) is as indicated in the Fig. 6.2. Note that $f(z)$ is equal to $\operatorname{Log} z + i\,6\pi$ on the negative real axis (dashed).

Remark 6.3   If (D, f) is a branch of the logarithm, then, for any fixed $k \in \mathbb{Z}$, so is (D, g), where $g(z) = f(z) + 2\pi k i$, for $z \in D$. Indeed, g is continuous, and $\exp g(z) = \exp f(z) \exp 2\pi k i = \exp f(z) = z$. The converse is true as we shall show next.

Theorem 6.2   *Suppose (D, f) and (D, g) are branches of the logarithm on the same domain D. Then there is $k \in \mathbb{Z}$ such that*

$$g(z) = f(z) + 2\pi i k$$

for all $z \in D$ (—the same k works for all $z \in D$).

Proof.   Set $h(z) = g(z) - f(z)$, for $z \in D$. Then

$$\exp h(z) = \exp\big(g(z) - f(z)\big) = \exp g(z) \exp\big(-f(z)\big)$$

$$= \frac{z}{\exp f(z)} = \frac{z}{z} = 1.$$

Hence, for each $z \in D$, there is some $k(z) \in \mathbb{Z}$ such that $h(z) = 2\pi i k(z)$.

Let $z_1, z_2 \in D$ be given. Since D is connected, we know that there is some path $\gamma : [a, b] \to \mathbb{C}$ in D joining z_1 to z_2. Now, h, and therefore k is continuous on D. Hence the map $t \mapsto k(\gamma(t))$ from $[a, b]$ into $\mathbb{Z}$ is continuous. It is therefore constant, by the Intermediate Value Theorem. Hence $k(\gamma(a)) = k(\gamma(b))$, that is, $k(z_1) = k(z_2)$ and we deduce that k is constant on D.   □

The next result tells us that any continuous choice of the logarithm is automatically differentiable and that its derivative is exactly what we would guess it to be, namely, $1/z$.

Theorem 6.3   *Let (D, f) be a branch of the logarithm. Then $f \in H(D)$ and $f'(z) = 1/z$, for all $z \in D$.*

Proof.   First we recall that $\exp f(z) = z$, for $z \in D$, means that $0 \notin D$. Also, if $z \neq w$, then $f(z) \neq f(w)$ (since otherwise, $z = e^{f(z)} = e^{f(w)} = w$).

Let $z, w \in D$, with $z \neq w$. Then

$$\frac{f(w) - f(z)}{w - z} = \frac{f(w) - f(z)}{e^{f(w)} - e^{f(z)}}$$
$$= \left(\frac{e^{f(w)} - e^{f(z)}}{f(w) - f(z)}\right)^{-1} \qquad (*)$$

Now, the continuity of f implies that if $w \to z$ then $f(w) \to f(z)$. Furthermore, for any $\zeta \in \mathbb{C}$,

$$\frac{e^{\xi} - e^{\zeta}}{\xi - \zeta} \to \exp' \zeta = e^{\zeta}$$

as $\xi \to \zeta$, with $\xi \neq \zeta$. Hence, as $w \to z$

$$(*) \quad \to \left(e^{f(z)}\right)^{-1} = \frac{1}{e^{f(z)}} = \frac{1}{z},$$

that is, $f \in H(D)$ and $f'(z) = \dfrac{1}{z}$, for any $z \in D$. $\qquad\square$

What's going on ? The notion of a complex logarithm is straightforward, but complicated by the fact that there is an infinite number of ways in which it can be defined. This is simply a consequence of the ambiguity in the choice of the polar angle, the argument of a complex number. To talk sensibly about a logarithm requires specifying some particular choice. This done, one then inquires about continuity considerations. This leads to the notion of branch of the logarithm where the region of definition of the logarithm is highlighted. It is not possible to make a continuous choice of logarithm in some domains — for example, in an annulus centred on the origin. The basic definition of a logarithm together with continuity is enough to imply its differentiability. Its derivative is $1/z$, as one might expect.

Once one has some notion of logarithm, it can then be used to construct complex powers.

As a corollary, we obtain an alternative proof of the above result on the uniqueness, up to an additive constant multiple of $2\pi i$, of the branch of the logarithm on a given domain.

Corollary 6.1 *Suppose (D, f) and (D, g) are branches of the logarithm on the same domain D. Then there exists some integer $k \in \mathbb{Z}$ such that $g(z) = f(z) + 2\pi k i$, for all $z \in D$.*

Proof. Since f and g are both logarithms, it follows that for each $z \in D$ there is an integer $k(z) \in \mathbb{Z}$ such that $h(z) = g(z) - f(z) = 2\pi k(z)i$. By the theorem, both f and g are differentiable on D and so, therefore, is h

and hence so is k. However, $k(z) \in \mathbb{R}$, for all $z \in D$ and so $k(z)$ is constant on the domain D.

Note that one could also argue that since f and g both have the same derivative on the domain D, namely $1/z$, then their difference h satisfies $h'(z) = 0$ on D. This means that h is constant on D and therefore of the form $2\pi ki$ for some fixed $k \in \mathbb{Z}$.   □

Example 6.7   Let S be the set $S = \{z : z = t\,e^{it}, \text{ where } t \in \mathbb{R}, \ t \geq 0\}$. We construct the branch (D, f) of the logarithm on the domain $D = \mathbb{C} \setminus S$ with $f(1) = \text{Log}\,1$.

The idea is to build on the values of $\text{Log}\,z$ by compensating for the jump in its value as z passes through the negative real axis.

Denote by $R_{-1}, R_0, R_1, \ldots$ the curved regions enclosed between the spiral S and the negative real axis, as shown in the Fig. 6.3, so that $D = \bigcup_{k=-1}^{\infty} R_k$. The section $(-\pi, 0)$ of the negative real axis is included in R_{-1}, the section $(-3\pi, -\pi)$ is included in R_0 and so on.

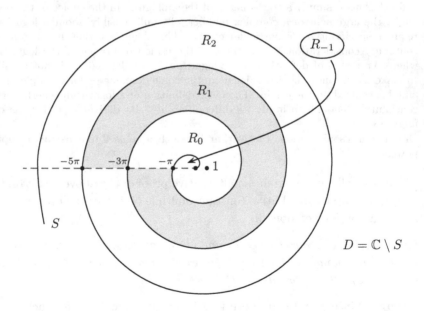

Fig. 6.3   The branch of logarithm on D with $f(1) = 0$.

The function f is defined on the domain D by setting

$$f(z) = \operatorname{Log} z + 2k\pi i, \quad \text{for } z \in R_k,\ k = -1, 0, 1, 2, \ldots.$$

Defined in this way, f is continuous on D and determines a branch of the logarithm. Furthermore, $f(1) = \operatorname{Log} 1$, since $1 \in R_0$. This uniquely specifies the branch (D, f).

Chapter 7

Complex Integration

7.1 Paths and Contours

In this chapter, we consider integration along a contour in the complex plane. Recall that a path in $\mathbb{C}$ is a continuous function $\gamma : [a, b] \to \mathbb{C}$, for some $a \le b$ in $\mathbb{R}$. The set of points $\{\, z : z = \gamma(t),\ a \le t \le b \,\}$ is the trace of γ which we denote by $\operatorname{tr} \gamma$.

Example 7.1 Let $\gamma_1(t) = e^{2\pi i t}$ for $0 \le t \le 1$. Then we find that

$$\operatorname{tr} \gamma_1 = \{\, z : |z| = 1 \,\} = \text{ unit circle.}$$

Now suppose that $\gamma_2(t) = e^{-4\pi i t}$ for $0 \le t \le 1$. We see that again

$$\operatorname{tr} \gamma_2 = \{\, z : |z| = 1 \,\} = \text{ unit circle.}$$

The paths γ_1 and γ_2 have the same trace but are different paths: the path γ_1 goes round the circle anticlockwise once, whereas γ_2 goes round twice in a clockwise sense.

Remark 7.1 The convention is to take the anticlockwise sense as being positive. This is consistent with the convention that the positive direction is that with increasing polar angle.

Definition 7.1 The path $\gamma : [a, b] \to \mathbb{C}$ is said to be closed if $\gamma(a) = \gamma(b)$. The path γ is said to be simple if it does not cross itself, that is, $\gamma(s) \ne \gamma(t)$ whenever $s \ne t$ with s, t in (a, b). (The possibility that $\gamma(a) = \gamma(b)$ is allowed.)

Example 7.2 The path γ_1 in the example above is a simple closed path, whereas γ_2 is closed but not simple. The path $\gamma(t) = e^{-5\pi i t}$, $0 \le t \le 1$, has the unit circle as its trace, but γ is neither closed nor simple.

Definition 7.2   Let $\gamma : [a, b] \to \mathbb{C}$ be a path. The reverse path $\widetilde{\gamma}$ is given by $\widetilde{\gamma} : [a, b] \to \mathbb{C}$ with $\widetilde{\gamma}(t) = \gamma(a + b - t)$.

Evidently, $\widetilde{\gamma}$ is "γ in the opposite direction". Note that the parametric range for $\widetilde{\gamma}$ is the same as that for γ, namely $[a, b]$. It is also clear that $\operatorname{tr} \widetilde{\gamma} = \operatorname{tr} \gamma$.

Definition 7.3   The path $\gamma : [a, b] \to \mathbb{C}$ is smooth if the derivative $\gamma'(t)$ exists and is continuous on $[a, b]$ (with right and left derivatives at a and b, respectively). In other words, if $\gamma(t) = x(t) + iy(t)$, then γ is smooth if and only if both x and y are differentiable (real functions of a real variable) and such that the derivatives $x'(t)$ and $y'(t)$ are continuous functions of the parameter $t \in [a, b]$.

A contour is a piecewise smooth path, that is, $\gamma : [a, b] \to \mathbb{C}$ is a contour if and only if there is a finite collection $a = a_0 < a_1 < \cdots < a_n = b$ (for some $n \in \mathbb{N}$) such that each subpath $\gamma : [a_{i-1}, a_i] \to \mathbb{C}$ is smooth, $1 \le i \le n$.

We write $\gamma = \gamma_1 + \gamma_2 + \cdots + \gamma_n$, where γ_i is the restriction of γ to the subinterval $[a_{i-1}, a_i]$.

Example 7.3   According to our (reasonable) definition, the path given by $t \mapsto \gamma(t) = \cos^3(2\pi t) + i \sin^3(2\pi t)$, for $0 \le t \le 1$, is smooth. Its trace is illustrated in the figure, Fig. 7.1.

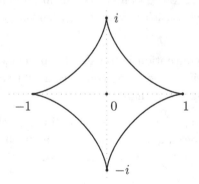

Fig. 7.1   A technically smooth path.

7.2 The Length of a Contour

Next, we wish to develop a means of formulating the length of a path. By way of motivation, suppose that $\gamma : [a, b] \to \mathbb{C}$ is a given smooth path. We can consider an inscribed polygon as an approximation to the path and its length will be an approximation to the length of the path. Suppose that the polygon has vertices $\{z_0, z_1, \ldots, z_n\}$. Then the length of such a polygonal path is the sum of the lengths of its straight line segments, namely, $\sum_{j=1}^{n} |z_j - z_{j-1}|$.

Suppose that the ends of the line segments correspond to values $t_0, t_1, \ldots, t_n$ of the path parameter, that is, $z_j = \gamma(t_j)$. Then

$$\sum_{j=1}^{n} |z_j - z_{j-1}| = \sum_{j=1}^{n} |\gamma(t_j) - \gamma(t_{j-1})|$$

$$= \sum_{j=1}^{n} \left| \frac{\gamma(t_j) - \gamma(t_{j-1})}{t_j - t_{j-1}} \right| (t_j - t_{j-1}),$$

with $a = t_0 < t_1 < \cdots < t_n = b$. Now, the quotient $\dfrac{\gamma(t_j) - \gamma(t_{j-1})}{t_j - t_{j-1}}$ can be considered an approximation to the derivative $\gamma'(t_j)$, in which case the sum on the right hand side becomes an approximation to an integral

$$\sum_{j=1}^{n} \left| \frac{\gamma(t_j) - \gamma(t_{j-1})}{t_j - t_{j-1}} \right| (t_j - t_{j-1}) \quad \overset{\text{approx}}{\rightsquigarrow} \quad \int_{a}^{b} |\gamma'(t)| \, dt.$$

Indeed, in view of the assumed smoothness of γ, one might expect that the left hand side converges to the right hand side as $\max_j (t_j - t_{j-1}) \to 0$. Rather than prove this here, we will just take this as motivation for our definition of the length of a path, as follows.

Definition 7.4 The length of the smooth path $\gamma : [a, b] \to \mathbb{C}$ is the non-negative real number $L(\gamma)$ given by the formula

$$L(\gamma) = \int_{a}^{b} |\gamma'(t)| \, dt.$$

If $\gamma = \gamma_1 + \gamma_2 + \cdots + \gamma_n$ is a contour (with each γ_j smooth), we define the length of γ to be $L(\gamma) = L(\gamma_1) + L(\gamma_2) + \cdots + L(\gamma_n)$.

Examples 7.4

(1) Let $\gamma(t) = e^{2\pi i t}$, for $0 \le t \le 1$. We calculate

$$L(\gamma) = \int_0^1 |\gamma'(t)|\, dt = \int_0^1 \left|2\pi i e^{2\pi i t}\right|\, dt = 2\pi \int_0^1 1\, dt = 2\pi\,.$$

(2) Let $\gamma(t) = e^{-4\pi i t}$, for $0 \le t \le 1$. Then $\gamma'(t) = -4\pi i e^{-4\pi i t}$, so that

$$L(\gamma) = \int_0^1 4\pi = 4\pi\,.$$

(3) Define the path γ by

$$\gamma(t) = \begin{cases} e^{2\pi i t}, & 0 \le t < 1 \\ 1, & 1 \le t < 2 \\ e^{-2\pi i t}, & 2 \le t \le 3. \end{cases}$$

Then γ is a contour and

$$\gamma'(t) = \begin{cases} 2\pi i e^{2\pi i t}, & 0 \le t < 1 \\ 0, & 1 < t < 2 \\ -2\pi i e^{-2\pi i t}, & 2 < t \le 3. \end{cases}$$

(At $t = 1$, the left derivative is $2\pi i$, the right derivative is 0 and at $t = 2$, the left derivative is 0, the right derivative is $-2\pi i$.) We find

$$L(\gamma) = \int_0^1 2\pi\, dt + \int_1^2 0\, dt + \int_2^3 2\pi\, dt = 4\pi\,.$$

(4) Suppose that $\gamma(t) = z_0 + t(z_1 - z_0)$, $0 \le t \le 1$. Then γ is the line segment $\gamma = [z_0, z_1]$. We have $\gamma'(t) = z_1 - z_0$ and so

$$L(\gamma) = \int_0^1 |z_1 - z_0|\, dt = |z_1 - z_0|\,,$$

as it should.

Remark 7.2   We see that every contour has a well-defined length—this is because of its good behaviour. In general, paths need not have a meaningful length. For example, let $\gamma : [0, 1] \to \mathbb{C}$ be the path given by

$$\gamma(t) = \begin{cases} (1 - t)\, e^{2\pi i/(1-t)}, & \text{for } 0 \le t < 1 \\ 0, & \text{for } t = 1. \end{cases}$$

The path γ starts at the point $\gamma(0) = 1$ and spirals counterclockwise in towards the origin (and encircles the origin an infinite number of times). Evidently, γ is continuous at any t with $0 \le t < 1$. Furthermore, for any $0 \le t < 1$, $|\gamma(t) - \gamma(1)| = (1 - t) \to 0$, as $t \uparrow 1$, and so we see that γ is also continuous at $t = 1$. Hence γ really is a path. However, for any $0 \le s < 1$, the length of γ "from 0 to s" is

$$\int_0^s |\gamma'(t)| \, dt = \int_0^s \left| \frac{2\pi i}{1 - t} - 1 \right| dt$$

$$\ge \int_0^s \frac{2\pi}{1 - t} \, dt = 2\pi \ln\left(\frac{1}{1 - s}\right) \to \infty$$

as s increases to 1. One could say that γ has infinite length.

7.3 Integration along a Contour

One way of looking at the integral of a real function of a real variable is to think of it as the area under its graph. This is approximated by the area of suitable rectangles (typically very thin and with height determined by the values of the function at the location of the rectangle).

$$\int_a^b f(x) \, dx = \text{ area under graph } \overset{\text{approx}}{=} \sum f(x_i) \Delta x_i$$

where $f(x_i) \Delta x_i$ is the area (height $\times$ base) of the rectangle with height $f(x_i)$ and width Δx_i. This formula makes sense for complex variables and complex-valued functions even though the original idea of area no longer does.

Suppose $\gamma : [a, b] \to \mathbb{C}$ is a smooth path. Let $a = t_0 < t_1 < \cdots < t_n = b$ and set $z_j = \gamma(t_j)$. With the preceding comments in mind, we consider

$$\sum f(z_j)(z_j - z_{j-1}) = \sum f(\gamma(t_j)) \frac{\gamma(t_j) - \gamma(t_{j-1})}{t_j - t_{j-1}} (t_j - t_{j-1})$$

$$\overset{\text{approx ?}}{=} \int_a^b f(\gamma(t)) \gamma'(t) \, dt$$

provided that the t_js are close together. It is possible to prove convergence of such sums to integrals (provided f is continuous), but we will simply take the right hand side as our definition of the complex contour integral.

Definition 7.5   Let $\gamma : [a, b] \to \mathbb{C}$ be smooth and f a continuous complex function on $\operatorname{tr} \gamma$. The contour integral of f along γ is defined to be

$$\int_{\gamma} f \equiv \int_{a}^{b} f(\gamma(t))\,\gamma'(t)\,dt.$$

Sometimes the notation $\int_{\gamma} f(z)\,dz$ is used. If we write $f = u + iv$ and $\gamma(t) = x(t) + iy(t)$, then $\int_{\gamma} f$ can be written as $X + iY$, where X and Y are real integrals,

$$X = \int_{a}^{b} \{u(x(t), y(t))\,x'(t) - v(x(t), y(t))\,y'(t)\}\,dt$$

and

$$Y = \int_{a}^{b} \{u(x(t), y(t))\,y'(t) + v(x(t), y(t))\,x'(t)\}\,dt.$$

For any contour γ, we define

$$\int_{\gamma} f = \int_{\gamma_1} f + \int_{\gamma_2} f + \cdots + \int_{\gamma_n} f$$

where $\gamma_1, \ldots, \gamma_n$ are the smooth parts of the contour γ.

Examples 7.5

(1) Let $\gamma(t) = e^{it}$ for $a \le t \le b$, and take $f(z) = \frac{1}{z}$. Then

$$\int_{\gamma} f \equiv \int_{\gamma} \frac{dz}{z}$$

$$= \int_{a}^{b} f(e^{it})\,ie^{it}\,dt$$

$$= \int_{a}^{b} \frac{1}{e^{it}}\,ie^{it}\,dt$$

$$= \int_{a}^{b} i\,dt = i(b - a).$$

In particular, with $a = 0$ and $b = 2\pi$, we find that $\int_{\gamma} \frac{dz}{z} = 2\pi i$.

(2) For given $a < b$ in $\mathbb{R}$, let $\gamma(t) = a + t(b - a)$, for $0 \le t \le 1$, that is, γ is just the line segment $[a, b]$ in $\mathbb{R}$. Then, for any f, continuous on $\operatorname{tr} \gamma$,

$$
\begin{aligned}
\int_\gamma f &= \int_0^1 f(\gamma(t))\, \gamma'(t)\, dt \\
&= \int_0^1 f(a + t(b - a))\, (b - a)\, dt \\
&= \int_a^b f(s)\, ds,
\end{aligned}
$$

by the change of variable $s = a + t(b - a)$. In particular, if f is real-valued, then $\int_{[a,b]} f$ is precisely the usual real integral $\int_a^b f(x)\, dx$.

Remark 7.3   This last observation has important consequences. Suppose that $\gamma = \gamma_1 + \cdots + \gamma_n$ is a contour where one of the smooth parts is a line segment on the real axis, $\gamma_k = [a, b]$, say, with $a < b \in \mathbb{R}$. Suppose that f is a continuous complex function on $\operatorname{tr} \gamma$ and that f is real on $\operatorname{tr} \gamma_k$, that is, f is real on the line segment $[a, b]$ on the real axis. We have

$$
\int_\gamma f = \int_{\gamma_1} f + \cdots + \int_{\gamma_k} f + \cdots + \int_{\gamma_n} f
$$

which expresses the real integral $\int_{\gamma_k} f = \int_a^b f(x)\, dx$ in terms of complex integrals. This may not seem much of an advance, but the point is that complex integrals can often be evaluated by general theory based only on rather general properties of the function f. This turns out to be a very powerful method for performing real integration.

Proposition 7.1   *For any contour γ and any functions, f, g, continuous on its trace $\operatorname{tr} \gamma$, we have*

(i) $\displaystyle \int_\gamma \alpha f + \beta g = \alpha \int_\gamma f + \beta \int_\gamma g, \quad \text{for all } \alpha,\, \beta \in \mathbb{C};$

(ii) $\displaystyle \int_\gamma f = -\int_{-\gamma} f.$

Proof.   The integral along a contour is given by the sum of the integrals along its smooth parts, so it is enough to consider the case of a smooth path $\gamma : [a, b] \to \mathbb{C}$. Then (i) is clear from the definition.

To prove (ii), we evaluate the left hand side.

$$\int_{\gamma} f = \int_{a}^{b} f(\tilde{\gamma}(t))\,\tilde{\gamma}'(t)\,dt$$

$$= -\int_{a}^{b} f(\gamma(a+b-t))\,\gamma'(a+b-t)\,dt$$

$$= \int_{b}^{a} f(\gamma(s))\,\gamma'(s)\,ds, \quad \text{putting } s = a+b-t,$$

$$= -\int_{a}^{b} f(\gamma(s))\,\gamma'(s)\,ds$$

$$= -\int_{\gamma} f,$$

as required.   □

The next result tells us that the integral along a path is somewhat insensitive to how we choose to parameterize the path.

Proposition 7.2   *The integral $\int_{\gamma} f$ is independent of the parametrization of the smooth path γ, i.e., if $\gamma : [a,b] \to \mathbb{C}$, $\varphi : [\alpha,\beta] \to [a,b]$ with $\varphi(\alpha) = a$, $\varphi(\beta) = b$, φ is differentiable and φ' is continuous, then*

$$\int_{\gamma} f = \int_{\widehat{\gamma}} f,$$

where $\widehat{\gamma}$ is the smooth path $\widehat{\gamma} = \gamma \circ \varphi : [\alpha,\beta] \to \mathbb{C}$.

Proof.   This is really just a change of variable. Let $f(\gamma(t))\,\gamma'(t) = u(t) + iv(t)$, where u and v are real, and let $U(s) = \int_{a}^{s} u(t)\,dt$ and $V(s) = \int_{a}^{s} v(t)\,dt$. Put $F(s) = U(s) + iV(s)$. Then $F : [a,b] \to \mathbb{C}$ and $F'(t) = U'(t) + iV'(t) = u(t) + iv(t) = f(\gamma(t))\,\gamma'(t)$. We have

$$\int_\gamma f = \int_a^b f(\gamma(t))\,\gamma'(t)\,dt = F(b) - F(a) = F(\varphi(\beta)) - F(\varphi(\alpha))$$
$$= U(\varphi(\beta)) - U(\varphi(\alpha)) + i\big(V(\varphi(\beta)) - V(\varphi(\alpha))\big)$$
$$= \int_\alpha^\beta U(\varphi(s))'\,ds + i\int_\alpha^\beta V(\varphi(s))'\,ds$$
$$= \int_\alpha^\beta U'(\varphi(s))\,\varphi'(s)\,ds + i\int_\alpha^\beta V'(\varphi(s))\,\varphi'(s)\,ds$$
$$= \int_\alpha^\beta F'(\varphi(s))\,\varphi'(s)\,ds = \int_\alpha^\beta f(\gamma(\varphi(s)))\,\gamma'(\varphi(s))\,\varphi'(s)\,ds$$
$$= \int_\alpha^\beta f(\gamma\circ\varphi(s))\,(\gamma\circ\varphi)'(s)\,ds = \int_{\gamma\circ\varphi} f\,,$$

as required. □

Example 7.6 For $0 \le t \le 1$, let $\gamma(t) = 1 + t(i-1)$, and let $\eta(t) = 1 + t^2(i-1)$. This corresponds to the reparametrization $\varphi(t) = t^2$, so that $\eta(t) = \gamma(\varphi(t))$. By way of illustration, let f be the function given by $z \mapsto f(z) = \operatorname{Re} z + 2\operatorname{Im} z$. Then

$$\int_\gamma f = \int_0^1 (1 - t + 2t)(i-1)\,dt$$
$$= i\int_0^1 (1+t)\,dt - \int_0^1 (1+t)\,dt.$$

On the other hand,

$$\int_\eta f = \int_0^1 (1 - t^2 + 2t^2)\,2t(i-1)\,dt$$
$$= i\int_0^1 (1+t^2)\,2t\,dt - \int_0^1 (1+t^2)\,2t\,dt$$
$$= i\int_0^1 (1+s)\,ds - \int_0^1 (1+s)\,ds, \quad \text{setting } s = t^2,$$

so $\int_\eta f = \int_\gamma f$, as it should. The actual value is $\frac{3}{2}(i-1)$.

7.4 Basic Estimate

The next result is a basic estimate which is used repeatedly. It is the complex analogue of the result from real analysis that an integral over an interval is bounded by any upper bound for (the modulus of) the integrand multiplied by the length of the interval. In the complex case, we consider contour integrals, and, in this case, the length of the interval is replaced by the length of the contour.

Theorem 7.1 (Basic Estimate) *Suppose that $\gamma : [a, b] \to \mathbb{C}$ is a smooth path and that f is a complex function, continuous on* $\operatorname{tr} \gamma$. *Then*

$$\left| \int_\gamma f \right| \leq \int_a^b |f(\gamma(t)) \, \gamma'(t)| \, dt.$$

In particular, for any contour γ, if $|f(\zeta)| \leq M$ for all $\zeta \in \operatorname{tr} \gamma$, then

$$\left| \int_\gamma f \right| \leq M \, L(\gamma).$$

Proof. If $\int_\gamma f = 0$, there is nothing to prove. So suppose that $\int_\gamma f \neq 0$, and let $\theta = \operatorname{Arg} \int_\gamma f$. Then $\int_\gamma f = e^{i\theta} \left| \int_\gamma f \right|$ and therefore

$$\left| \int_\gamma f \right| = e^{-i\theta} \int_\gamma f$$

$$= \int_a^b e^{-i\theta} f(\gamma(t)) \, \gamma'(t) \, dt$$

$$= \int_a^b \operatorname{Re}\!\left(e^{-i\theta} f(\gamma(t))\gamma'(t)\right) dt + i \underbrace{\int_a^b \operatorname{Im}\!\left(e^{-i\theta} f(\gamma(t))\gamma'(t)\right) dt}_{=\,0 \text{ since lhs is real}}$$

$$\leq \int_a^b \left| e^{-i\theta} f(\gamma(t))\gamma'(t) \right| dt, \quad \text{since } \operatorname{Re} w \leq |w|, \text{ for any } w \in \mathbb{C},$$

$$= \int_a^b |f(\gamma(t))| \, |\gamma'(t)| \, dt,$$

as claimed. Now, if $|f(\zeta)| \leq M$ for all $\zeta \in \operatorname{tr}\gamma$, it follows that $|f(\gamma(t))| \leq M$ for all $a \leq t \leq b$ and so

$$
\begin{aligned}
\left| \int_\gamma f \right| &\leq \int_a^b |f(\gamma(t))| \, |\gamma'(t)| \, dt \\
&\leq \int_a^b M \, |\gamma'(t)| \, dt \\
&= M \, L(\gamma).
\end{aligned}
$$

The result for a contour follows by summing over its smooth subpaths;

$$
\begin{aligned}
\left| \int_\gamma f \right| &= \left| \int_{\gamma_1 + \cdots + \gamma_n} f \right| \\
&= \left| \int_{\gamma_1} f + \cdots + \int_{\gamma_n} f \right| \\
&\leq \left| \int_{\gamma_1} f \right| + \cdots + \left| \int_{\gamma_n} f \right| \\
&\leq M \left(L(\gamma_1) + \cdots + L(\gamma_n) \right) \\
&= M \, L(\gamma),
\end{aligned}
$$

as required. □

7.5 Fundamental Theorem of Calculus

The next result is a version of the Fundamental Theorem of Calculus in the setting of contour integration.

Theorem 7.2   *Suppose that f is differentiable and that its derivative f' is continuous on the trace $\operatorname{tr}\gamma$ of a contour $\gamma : [a, b] \to \mathbb{C}$. Then*

$$
\int_\gamma f' = f(\gamma(b)) - f(\gamma(a)).
$$

Proof. Suppose first that γ is smooth. Then we have

$$\int_\gamma f' = \int_a^b f'(\gamma(t))\,\gamma'(t)\,dt$$

$$= \int_a^b \frac{d}{dt}\big(f(\gamma(t))\big)\,dt$$

$$= \int_a^b \big(U'(t) + iV'(t)\big)\,dt$$

where $U(t) = \operatorname{Re} f(\gamma(t))$ and $V(t) = \operatorname{Im} f(\gamma(t))$

$$= U(b) - U(a) + i(V(b) - V(a))$$

$$= f(\gamma(b)) - f(\gamma(a)).$$

For the general case, suppose that γ is the contour $\gamma = \gamma_1 + \cdots + \gamma_k$, where each $\gamma_j = \gamma : [t_{j-1}, t_j] \to \mathbb{C}$, $j = 1, \ldots, k$, is a smooth path (the restriction of γ to the subinterval $[t_{j-1}, t_j]$) for suitable $a = t_0 < t_1 < \cdots < t_k = b$. Then, as above, we find that

$$\int_\gamma f' = \sum_{j=1}^k \int_{\gamma_j} f'$$

$$= \sum_{j=1}^k \big(f(\gamma(t_j)) - f(\gamma(t_{j-1})) \big)$$

$$= f(\gamma(b)) - f(\gamma(a)),$$

since the sum telescopes. $\qquad\square$

As a corollary, we recover a familiar result.

Corollary 7.1 *Suppose that $f \in H(D)$ and that $f' = 0$ on D. Then f is constant on D.*

Proof. For any pair of points, z_0 and z_1, in D, let $\gamma : [a, b] \to \mathbb{C}$ be a polygon in D joining z_0 to z_1. By hypothesis, $f' = 0$ on D and so f' satisfies the hypothesis of the preceding theorem. Hence

$$\int_\gamma f' = f(\gamma(b)) - f(\gamma(a)) = f(z_1) - f(z_0).$$

But $f' = 0$ which implies that $\int_\gamma f' = 0$. It follows that $f(z_0) = f(z_1)$ and we conclude that f is constant on D. $\qquad\square$

7.6 Primitives

Definition 7.6   Let D be a domain and suppose $f : D \to \mathbb{C}$ is continuous. A map $F : D \to \mathbb{C}$ such that $F' = f$ on D is said to be a primitive for the function f on D.

Corollary 7.2   *If F is a primitive for f on D, then*

$$\int_\gamma f = F(\gamma(b)) - F(\gamma(a))$$

for any contour lying in D. In particular, if γ is closed and f has a primitive in D, then

$$\int_\gamma f = 0$$

provided $\operatorname{tr}\gamma \subset D$.

Proof.   By theorem 7.2, we see immediately that

$$\int_\gamma f = \int_\gamma F' = F(\gamma(b)) - F(\gamma(a)).$$

In particular, $\gamma(a) = \gamma(b)$ if γ is closed and so $\int_\gamma f = 0$. $\square$

Remark 7.4   The existence of a primitive depends not only on the form of the function, but also on the domain in question. For example, we have seen that $\operatorname{Log} z$ has derivative $\frac{1}{z}$ on the cut-plane. In other words, $\operatorname{Log} z$ is a primitive for $\frac{1}{z}$ on the cut-plane $\mathbb{C} \setminus \{ z : z + |z| = 0 \}$. However, let $\gamma(t) = e^{2\pi i t}$, $0 \le t \le 1$, and let D be any domain containing $\operatorname{tr}\gamma$ (but excluding 0). Then $\frac{1}{z}$ has *no* primitive on D.

This follows from the observation that γ is closed but $\int_\gamma \frac{1}{z} = 2\pi i \ne 0$. It follows that there is no branch of the logarithm on such D. Indeed, any branch of the logarithm would be a primitive for the function $\frac{1}{z}$, and no such primitive exists in this case.

Example 7.7   For any $n \in \mathbb{Z}$, with $n \ne -1$, the function $F(z) = z^{n+1}/(n+1)$ is a primitive for z^n on $\mathbb{C}$ (or on $\mathbb{C} \setminus \{0\}$, if $n < -1$). It follows that $\int_\gamma z^n \, dz = 0$ for any closed contour γ (lying in $\mathbb{C} \setminus \{0\}$, if $n < -1$).

Furthermore, considering the line segment $[z_0, z_1]$, we find that

$$\int_{[z_0, z_1]} z^n dz = \frac{1}{n+1}\left(z_1^{n+1} - z_0^{n+1} \right)$$

for $n \in \mathbb{Z}$, $n \neq -1$. In particular,

$$\int_{[z_0, z_1]} dz = z_1 - z_0 \,.$$

Of course, if $n < -1$ in the above, then we must assume that $0 \notin \mathrm{tr}\,\gamma$, i.e., that the line segment $[z_0, z_1]$ does not pass through the origin.

Example 7.8   Let $f(z) = a_0 + a_1 z + \cdots + a_n z^n$ be a polynomial. Then the coefficients a_k satisfy

$$a_k = \frac{1}{2\pi i} \int_\gamma \frac{f(z)}{z^{k+1}} \, dz \,,$$

where $\gamma : [0, 1] \to \mathbb{C}$ is the circle $\gamma(t) = r\, e^{2\pi i t}$, for $0 \leq t \leq 1$ and any $r > 0$.

This follows from the fact that z^n has a primitive in $\mathbb{C} \setminus \{0\}$ for any $n \in \mathbb{Z}$ with $n \neq -1$, together with the fact that $\int_\gamma dz/z = 2\pi i$. This type of integral formula is very important and will crop up again in a more general setting.

Example 7.9   Let $\gamma : [0, 1] \to \mathbb{C}$ be the path $\gamma(t) = 1 + t + i \sin\!\left(\frac{5}{2}\pi t\right)$.

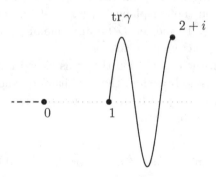

Fig. 7.2   The trace of the path γ.

We wish to determine $\displaystyle\int_\gamma \frac{1}{z} \, dz \,.$

To do this, we simply note that the function $1/z$ has primitive $\operatorname{Log} z$ in the star-domain $\mathbb{C} \setminus \{z + |z| = 0\}$, so

$$\int_\gamma \frac{dz}{z} = \operatorname{Log} \gamma(1) - \operatorname{Log} \gamma(0)$$
$$= \operatorname{Log}(2 + i) - \operatorname{Log} 1$$
$$= \operatorname{Log}(2 + i)$$
$$\left(= \tfrac{1}{2}\ln 5 + i \tan^{-1}(\tfrac{1}{2}) \right).$$

This is somewhat easier than trying to directly evaluate $\int_0^1 f(\gamma(t))\,\gamma'(t)\,dt$.

To plot the trace of γ, we note that the point $z = x + iy \leftrightarrow (x, y)$ belongs to $\operatorname{tr} \gamma$ provided $x \in [1, 2]$ and $y = \sin\left(\tfrac{5}{2}\pi(x - 1)\right)$.

Chapter 8

Cauchy's Theorem

8.1 Cauchy's Theorem for a Triangle

In this chapter, we shall discuss the central theorem of complex analysis, Cauchy's Theorem. This involves integrals around triangles, so first we define exactly what we mean.

Definition 8.1 A triangle T with vertices $\{z_1, z_2, z_3\}$ is the set

$$T = \{\, z \in \mathbb{C} : z = \alpha z_1 + \beta z_2 + \gamma z_3 \quad \text{with } \alpha, \beta, \gamma \geq 0 \text{ and } \alpha + \beta + \gamma = 1. \,\}$$

Thus, by triangle, we mean the interior as well as the edges. Note that T is closed. To avoid unnecessary extra notation, we shall denote the integral around the closed contour $[z_1, z_2] + [z_2, z_3] + [z_3, z_1]$, i.e., the integral around the sides of the triangle, by $\int_{\partial T}$.

Denote the length of this contour by $L(\partial T)$; thus,

$$L(\partial T) = L([z_1, z_2]) + L([z_2, z_3]) + L([z_3, z_1])$$
$$= |z_2 - z_1| + |z_3 - z_2| + |z_1 - z_3|.$$

A moment's consideration reveals that $\operatorname{diam} T \leq L(\partial T)$. In fact, $\operatorname{diam} T$ is equal to the length of the longest side of the triangle T.

Theorem 8.1 (Cauchy's Theorem for a Triangle) *For any given domain D and any function $f \in H(D)$,*

$$\int_{\partial T} f = 0$$

for any triangle T (wholly) contained in D.

Proof.   We emphasize that the triangle T (where we include its whole convex interior) must be contained in D. It is *not* enough that just the three sides belong to D, which could happen if D has "holes".

Bisect each of the sides of the triangle T to obtain four smaller triangles, $T_1, \ldots, T_4$, as shown in Fig. 8.1.

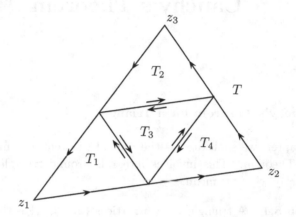

Fig. 8.1   Construct 4 congruent triangles by bisecting the sides of T.

Then

$$\int_{\partial T} f = \int_{\partial T_1} f + \int_{\partial T_2} f + \int_{\partial T_3} f + \int_{\partial T_4} f,$$

since each integral around a triangle is given as the sum of three integrals along line segments, and the integrals along those line segments in the interior of T cancel out. It follows that

$$\left| \int_{\partial T} f \right| \leq \left| \int_{\partial T_1} f \right| + \left| \int_{\partial T_2} f \right| + \left| \int_{\partial T_3} f \right| + \left| \int_{\partial T_4} f \right|. \qquad (*)$$

Let $\widehat{T}_1$ be one of $T_1, \ldots, T_4$ giving the maximum contribution to the right hand side of the inequality $(*)$, so that

$$\left| \int_{\partial T} f \right| \leq 4 \left| \int_{\partial T_1} f \right|.$$

We have that $\widehat{T}_1 \subset T$ and $L(\partial \widehat{T}_1) = \frac{1}{2} L(\partial T)$, since each side of $\widehat{T}_1$ is one half that of the corresponding side of T.

We now repeat this procedure cutting $\widehat{T}_1$ into 4 smaller triangles to obtain some triangle $\widehat{T}_2$ with $\widehat{T}_2 \subset \widehat{T}_1$, $L(\partial \widehat{T}_2) = \frac{1}{2}L(\partial \widehat{T}_1) = \frac{1}{2^2}L(\partial T)$ and

$$\left| \int_{\partial T} f \right| \leq 4 \left| \int_{\partial T_1} f \right| \leq 4^2 \left| \int_{\partial T_2} f \right|.$$

We can do this again for $\widehat{T}_2$, and so on. In this way, we obtain a nested sequence of triangles $T \supset \widehat{T}_1 \supset \widehat{T}_2 \supset \ldots$ such that $L(\partial T) = 2^n L(\partial \widehat{T}_n)$ and

$$\left| \int_{\partial T} f \right| \leq 4 \left| \int_{\partial T_1} f \right| \leq \cdots \leq 4^n \left| \int_{\partial T_n} f \right|.$$

Now, $\operatorname{diam} \widehat{T}_n = 2^{-n} \operatorname{diam} T \to 0$, as $n \to \infty$. Furthermore, each $\widehat{T}_n$ is closed, so we can apply Cantor's Theorem, theorem 3.3, to conclude that there is some z_0 such that $\{z_0\} = \bigcap_n \widehat{T}_n$.

Next, we note that $z_0 \in D$ and so f is differentiable at z_0 because f is analytic in D. Define the function τ on D by the formula

$$\tau(z) = \begin{cases} \dfrac{f(z) - f(z_0)}{z - z_0} - f'(z_0), & \text{for } z \neq z_0 \\ 0, & \text{for } z = z_0. \end{cases}$$

Then τ is continuous on D and, in particular, $\tau(z) \to 0$ as $z \to z_0$. We can express f in terms of τ as follows,

$$f(z) = f(z_0) + (z - z_0)f'(z_0) + (z - z_0)\tau(z), \quad \text{for } z \in D.$$

Hence

$$\int_{\partial T_n} f = \int_{\partial T_n} \left\{ f(z_0) + (z - z_0)f'(z_0) + (z - z_0)\tau(z) \right\} dz$$

$$= 0 + 0 + \int_{\partial T_n} (z - z_0)\tau(z)\, dz,$$

since $\int_{\partial T_n} dz = \int_{\partial T_n} (z - z_0)\, dz = 0$ because the integrands have primitives (on the whole of $\mathbb{C}$).

We must estimate the (modulus of the) third integral in the equation above. (For large n, it is an integral around a very small triangle, located around the point z_0, of a function which is close to zero in this region. We therefore expect this integral to have small modulus. However, the earlier estimate for $\left| \int_{\partial T} f \right|$ in terms of $\left| \int_{\partial T_n} f \right|$ involves the factor 4^n, so we must pay attention to the details here.)

Let $\varepsilon > 0$ be given. Since $\tau(z) \to 0$ as $z \to z_0$, we know that there is some $\delta > 0$ such that $|\tau(z)| < \varepsilon$ whenever $|z - z_0| < \delta$. Furthermore, we know that $\operatorname{diam} \widehat{T}_n = 2^{-n} \operatorname{diam} T \to 0$ as $n \to \infty$, and so we may choose n so that $\operatorname{diam} \widehat{T}_n < \delta$.

Then any for $z \in \partial \widehat{T}_n$, the boundary of the triangle $\widehat{T}_n$, it follows that $|z - z_0| \leq \operatorname{diam} \widehat{T}_n < \delta$, since $z_0 \in \widehat{T}_n$. Hence $|\tau(z)| < \varepsilon$ and therefore

$$|(z - z_0)\tau(z)| \leq \operatorname{diam} \widehat{T}_n \; \varepsilon$$

for any $z \in \partial \widehat{T}_n$. It follows that

$$\left| \int_{\partial T_n} (z - z_0)\tau(z)\, dz \right| \leq \operatorname{diam} \widehat{T}_n \; \varepsilon \, L(\partial \widehat{T}_n)$$

$$= \frac{\varepsilon \, \operatorname{diam} T \, L(\partial T)}{4^n}$$

and therefore

$$\left| \int_{\partial T_n} f \right| = \left| \int_{\partial T_n} (z - z_0)\tau(z)\, dz \right| \leq \frac{\varepsilon \, \operatorname{diam} T \, L(\partial T)}{4^n}.$$

Hence, finally, we have

$$\left| \int_{\partial T} f \right| \leq 4^n \left| \int_{\partial T_n} f \right|$$

$$\leq 4^n \, \frac{\varepsilon \, \operatorname{diam} T \, L(\partial T)}{4^n}$$

$$= \varepsilon \, \operatorname{diam} T \; L(\partial T).$$

This holds for any $\varepsilon > 0$ and so we conclude that $\int_{\partial T} f = 0$. $\square$

Theorem 8.2   *Let D be a star-domain. Then every function analytic in D has a primitive, i.e., if $f \in H(D)$, then there is $F \in H(D)$ such that $F' = f$ on D.*

Proof.   Let z_0 be a star-centre for D. Define F on D by

$$F(z) = \int_{[z_0, z]} f$$

for $z \in D$. Note that F is well-defined since $[z_0, z] \subset D$. We claim that F is differentiable and that $F' = f$ everywhere in D. To show this, let $z \in D$ be given. Since D is open, there is $r > 0$ such that $D(z, r) \subseteq D$. We wish to show that

$$\frac{F(z + \zeta) - F(z)}{\zeta} - f(z) \to 0$$

as $\zeta \to 0$. We are assured that $F(z + \zeta)$ is defined for all ζ with $|\zeta| < r$, since $D(z, r) \subseteq D$, as noted above. Suppose that $|\zeta| < r$ from now on.

We will appeal to Cauchy's Theorem to rewrite $F(z + \zeta) - F(z)$ in another form. Indeed,

$$F(z + \zeta) - F(z) = \int_{[z_0, z+\zeta]} f - \int_{[z_0, z]} f.$$

Let T denote the triangle with vertices z_0, $z + \zeta$ and z. Since z_0 is a star-centre for D, and since $[z + \zeta, z]$, the line segment from $z + \zeta$ to z, lies in $D(z, r) \subseteq D$, it follows that $T \subset D$. Indeed, any point w on the line segment $[z + \zeta, z]$ lies in D and so, therefore, does the line segment $[z_0, w]$. By varying w, we exhaust the triangle T. Now, by Cauchy's Theorem, $\int_{\partial T} f = 0$. However, the contour integral around a triangle is equal (by definition) to the sum of the integrals along its sides. Hence, we have

$$\int_{[z_0, z+\zeta]} f + \int_{[z+\zeta, z]} f + \int_{[z, z_0]} f = 0.$$

Reversing the direction of the contour is equivalent to a change in sign of the integral, and therefore we may rewrite the above equation as

$$\int_{[z_0, z+\zeta]} f - \int_{[z_0, z]} f = -\int_{[z+\zeta, z]} f = \int_{[z, z+\zeta]} f.$$

In terms of F, this becomes

$$F(z + \zeta) - F(z) = \int_{[z, z+\zeta]} f(\xi) \, d\xi.$$

Hence

$$\frac{F(z + \zeta) - F(z)}{\zeta} - f(z) = \frac{1}{\zeta} \int_{[z, z+\zeta]} f(\xi) \, d\xi - f(z)$$

$$= \int_{[z, z+\zeta]} \frac{f(\xi) - f(z)}{\zeta} \, d\xi$$

since $f(z) \int_{[z, z+\zeta]} d\xi = f(z) \zeta$. It remains to estimate this last integral. To do this, we use the continuity of f at z.

Let $\varepsilon > 0$ be given. Then there is $\delta > 0$ such that $|f(\xi) - f(z)| < \varepsilon$ whenever $|\xi - z| < \delta$. It follows that if $|\zeta| < \min\{\delta, r\}$, then $|\xi - z| < \delta$ for every $\xi \in [z, z + \zeta]$ so that $|f(\xi) - f(z)| < \varepsilon$ for such ξ. Therefore

$$\left| \frac{F(z + \zeta) - F(z)}{\zeta} - f(z) \right| = \frac{1}{|\zeta|} \left| \int_{[z, z+\zeta]} \big(f(\xi) - f(z)\big) \, d\xi \right|$$

$$\leq \frac{1}{|\zeta|} \, \varepsilon \, L([z, z + \zeta])$$

$$= \frac{1}{|\zeta|} \, \varepsilon \, |\zeta|$$

$$= \varepsilon$$

and the result follows.   □

What's going on ?   The method is constructive in that an explicit formula for a primitive is given, rather than it merely being argued to exist. To achieve this, use is made of the existence of a star-centre for D. The fact that z_0 is a star-centre means that every line segment $[z_0, z]$ lies in D whenever z does. This, together with the continuity of f ensures that F is well-defined. The fact that F actually is a primitive for f is a consequence of the continuity of f and the fact that the integral of f around any triangle is zero. Of course, these two facts are consequences of the (assumed) analyticity of f in D, but the proof that F is a primitive for f carries through if the hypothesis "f is analytic in D" is directly replaced by "f is continuous in D and its integral around any triangle in D is zero". Stating the theorem this way might seem rather artificial or contrived, but we will see later (see Morera's Theorem, theorem 8.8) that it turns out to be quite relevant.

As a consequence of this last result, we show that if f is does not vanish on a star-domain, then f has an analytic logarithm and also an analytic n^{th}-root.

Theorem 8.3   *Suppose that f is analytic in the star-domain D and that $f(z) \neq 0$ for any $z \in D$. Then there is $g \in H(D)$ such that $f(z) = \exp g(z)$ for all $z \in D$.*

In particular, for any $n \in \mathbb{N}$ there is $h \in H(D)$ such that $h^n = f$ in D.

Proof.   Since f does not vanish in D, the function f'/f is analytic in D. But D is star-like and so, by Theorem 8.2, f'/f has a primitive there, say $F \in H(D)$. Then $F' = f'/f$ in D. (Experience from calculus suggests that F might be a good candidate for a logarithm of f. This is almost correct.) Now let $\psi(z) = f(z) \, e^{-F(z)}$ for $z \in D$. Then we calculate the derivative

$$\psi'(z) = f'(z)e^{-F(z)} - F'(z)f(z)e^{-F(z)} = 0$$

because $F'(z)f(z) = f'(z)$. Since D is connected, it follows that ψ is constant on D.

Fix $z_0 \in D$. Then

$$f(z)e^{-F(z)} = \psi(z) = \psi(z_0) = f(z_0)e^{-F(z_0)}$$

for any $z \in D$. Rearranging and using $f(z_0) = e^{\text{Log } f(z_0)}$, we get

$$f(z) = e^{F(z)-F(z_0)+\text{Log } f(z_0)}.$$

Let $g(z) = F(z) - F(z_0) + \text{Log } f(z_0)$. Then $g \in H(D)$ and obeys

$$e^{g(z)} = f(z),$$

as required.

It is now easy to construct an analytic n^{th}-root of f in D. Indeed, for any $n \in \mathbb{N}$, set $h(z) = \exp\left(\frac{1}{n} g(z)\right)$. Then

$$h^n(z) = \exp(g(z)) = f(z)$$

and the proof is complete. $\qquad\qquad\qquad\qquad\qquad\qquad\qquad\qquad\qquad\qquad$ □

8.2 Cauchy's Theorem for Star-Domains

Theorem 8.4 (Cauchy's Theorem for Star-Domains) *Let f be analytic in a star-domain D. Then $\int_\gamma f = 0$ for any closed contour γ with $\text{tr }\gamma \subset D$. Moreover, if φ and ψ are contours in D with the same initial point z_0 and the same final point z_1, then $\int_\varphi f = \int_\psi f$.*

Proof. By theorem 8.2, any $f \in H(D)$ has a primitive on D, that is, there is $F \in H(D)$ with $F' = f$ on D. Let $\gamma : [a, b] \to \mathbb{C}$ be any contour in D. Then

$$\int_\gamma f = \int_\gamma F' = F(\gamma(b)) - F(\gamma(a)) = 0,$$

if γ is closed. For the last part, as above, we have

$$\int_\varphi f = F(z_1) - F(z_0) = \int_\psi f,$$

as required. $\qquad\qquad\qquad\qquad\qquad\qquad\qquad\qquad\qquad\qquad\qquad\qquad\qquad$ □

This central result can be generalized to more general domains, but troubles may arise when D has holes.

Example 8.1   Suppose that $f \in H(D)$ and that γ is a contour lying in D, as indicated in Fig. 8.2.

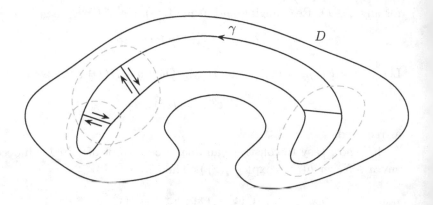

Fig. 8.2 $\int_\gamma f$ may still vanish in certain non star-like domains.

We see that $\int_\gamma f = 0$ even though D is clearly not star-like. The reason is that we can put in cross-cuts and use the fact that each sub-contour *is* in a star-domain in which f is analytic. The contributions to the integral from the extra cross-cuts cancel out because they are eventually traversed in both directions. In other words, we may sometimes be able to piece together overlapping star-domains to get a non star-like domain for which the theorem nevertheless remains valid. This kind of argument is often done "by inspection", that is to say, the precise way the contour is cut up will generally depend on the particular case in hand.

Example 8.2   It is important to realize that this trick cannot be used if D has holes and the contour γ goes around such a hole. For example, suppose that the domain D is $D = \mathbb{C} \setminus \{0\}$, the punctured plane. Let γ be the circle around 0 given by $\gamma(t) = e^{2\pi i t}$, for $0 \le t \le 1$, and let $f(z) = \frac{1}{z}$. Then we know that

$$\int_\gamma f = \int_\gamma \frac{1}{z}\, dz = 2\pi i$$

and so, in this case, $\int_\gamma f \ne 0$.

Example 8.3 By way of illustration, we evaluate the contour integral

$$\int_\Gamma \frac{z^2+1}{z^3}\,dz\,,$$

where Γ is the simple closed contour (polygon, in the counter clockwise sense) whose trace is the square $ABCD$ with corners at $A = -1 - i$, $B = 1 - i$, $C = 1 + i$ and $D = -1 + i$.

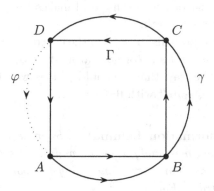

Fig. 8.3 Applying Cauchy's Theorem.

The function $1/z^3$ has a primitive on $\mathbb{C} \setminus \{0\}$ and so its integral around any closed contour (not passing through the point 0) vanishes. The function $1/z$ is analytic in $\mathbb{C} \setminus \{0\}$. The domain $\mathbb{C} \setminus \{z + |z| = 0\}$ is star-like and so the integral of $1/z$ around the three sides of the square $A \to B \to C \to D$ is the same as that from A to D along the circular arc $\gamma(t) = \sqrt{2}e^{it}$, $-3\pi/4 \le t \le 3\pi/4$, namely

$$\int_\gamma \frac{dz}{z} = \int_{-3\pi/4}^{3\pi/4} \frac{\sqrt{2}\,ie^{it}}{\sqrt{2}e^{it}}\,dt = \frac{3\pi i}{2}.$$

Next, we note that $\{z : \operatorname{Re} z < 0\}$ is star-like and so the integral of $1/z$ along the side $D \to A$ of the square is equal to that from D to A along the arc (dotted in Fig. 8.3) $\varphi(t) = \sqrt{2}e^{it}$, $3\pi/4 \le t \le 5\pi/4$, namely

$$\int_\varphi \frac{dz}{z} = \int_{3\pi/4}^{5\pi/4} \frac{\sqrt{2}\,ie^{it}}{\sqrt{2}e^{it}}\,dt = \frac{\pi i}{2}.$$

Adding, we find

$$\int_\Gamma \frac{z^2+1}{z^3}\,dz = \int_\Gamma \frac{1}{z}\,dz + 0 = \int_\gamma \frac{dz}{z} + \int_\varphi \frac{dz}{z} = 2\pi i.$$

8.3   Deformation Lemma

Before proceeding, it is convenient to introduce some terminology. In order to avoid some gratuitous circumlocution, let us agree to call a contour γ a simple circular curve, or just a circle, if it has the form $\gamma(t) = \zeta + \rho\, e^{2\pi it}$, $0 \leq t \leq 1$, for some $\zeta \in \mathbb{C}$ and $\rho > 0$. Here, tr γ is a circle with radius ρ and with centre at the point ζ in $\mathbb{C}$. The path is traced out in the usual positive (i.e., counter-clockwise) sense and makes just one single circuit. γ is a simple closed smooth path.

If w is any point whose distance from the centre ζ is smaller than ρ, then we shall say that γ encircles w (or goes around w) or that w lies inside the circle γ. Note, by the way, that we could just as well have parameterized this circle as $\gamma(t) = \zeta + \rho\, e^{it}$ with $0 \leq t \leq 2\pi$.

Lemma 8.1 (Deformation Lemma)   *Suppose that ζ belongs to the disc $D(a, R)$ and that g is analytic in the punctured set $D(a, R) \setminus \{\zeta\}$. Let γ and Γ be simple circular curves encircling the point ζ and lying inside $D(a, R)$, as indicated in Fig. 8.4. Then*

$$\int_\gamma g = \int_\Gamma g\,.$$

In particular, if f is analytic in $D(a, R)$, then

$$\int_\gamma \frac{f(w)}{w - \zeta}\, dw = \int_\Gamma \frac{f(w)}{w - \zeta}\, dw.$$

Proof.   The hypotheses are that $\gamma(t) = \zeta_1 + \rho_1 e^{2\pi it}$ and $\Gamma(t) = \zeta_2 + \rho_2 e^{2\pi it}$, for $0 \leq t \leq 1$, for some $\zeta_1, \zeta_2 \in C$ and for some $\rho_1 > 0$ and $\rho_2 > 0$. Furthermore, tr $\gamma \subset D(a, R)$, tr $\Gamma \subset D(a, R)$, $|\zeta - \zeta_1| < \rho_1$ and $|\zeta - \zeta_2| < \rho_2$. Without loss of generality, we may assume that the circle γ has a small radius and so is inside Γ, as illustrated in Fig. 8.4. (If not, we just show that each integral is equal to that around one such small circle and so are equal to each other.) The idea is to put in cross-cuts and apply Cauchy's Theorem in suitable resultant star-like regions.

With the notation of Fig. 8.4, we see that

$$\int_{A+B+C+D+A'+B'+C'+D'} g = \int_{\Gamma+\gamma} g$$

since the contributions to the contour integral along the parts B and B' cancel out, as do those along D and D'.

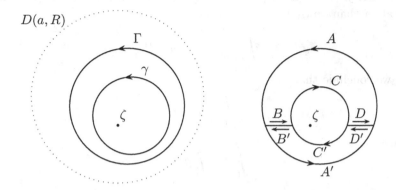

Fig. 8.4 Put in cross-cuts.

Now, $A + B + C + D$ is a closed curve which is contained in a star-domain $D_1 \subseteq D(a, R) \setminus \{\zeta\}$, as shown in Fig. 8.5, and g is analytic in D_1. (We can take D_1 to be the star-domain $D(a, R) \cap (\mathbb{C} \setminus L_1)$, where L_1 is any straight line (ray) from ζ which does not cut $A + B + C + D$, as shown in Fig. 8.5.) Hence, by Cauchy's Theorem, theorem 8.4,

$$\int_{A+B+C+D} g = 0 .$$

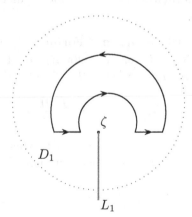

Fig. 8.5 There is an enclosing star-domain.

Similarly, (replacing L_1 by another suitable straight line cut (this time up rather than down))

$$\int_{A'+B'+C'+D'} g = 0$$

and we conclude that

$$\int_{\Gamma+\gamma} g = 0.$$

Hence

$$\int_\Gamma g - \int_\gamma g = 0,$$

as required. For the last (important) part, take $g(w) = f(w)/(w - \zeta)$. □

Remark 8.1 This says that the contour γ can be moved or "deformed" into the contour Γ without changing the value of the integral, provided the change is through a region of analyticity of g.

As can be seen from the proof of the Deformation Lemma, the disc $D(a, R)$ could be replaced by a more general shaped domain, and the closed contours γ and Γ need not be circular. However, the case with circles is of particular interest as we will see later.

8.4 Cauchy's Integral Formula

We now use the Deformation Lemma to obtain another of the basic results of complex analysis.

Theorem 8.5 (Cauchy's Integral Formula) *Suppose f is analytic in the open disc $D(a, R)$ and that $z_0 \in D(a, R)$. Let Γ be any simple circular contour around z_0 and lying in $D(a, R)$. Then*

$$f(z_0) = \frac{1}{2\pi i} \int_\Gamma \frac{f(w)}{w - z_0}\, dw.$$

Proof. Let $\gamma(t) = z_0 + \rho\, e^{2\pi i t}$, $0 \le t \le 1$, where $\rho > 0$ is chosen small enough that $\operatorname{tr}\gamma \subset D(a, R)$. Using the Deformation Lemma, we have

$$\int_\Gamma \frac{f(w)}{w - z_0}\, dw = \int_\gamma \frac{f(w)}{w - z_0}\, dw$$

$$= \int_\gamma \frac{f(w) - f(z_0)}{w - z_0}\, dw + \int_\gamma \frac{f(z_0)}{w - z_0}\, dw.$$

Now,

$$\int_\gamma \frac{f(z_0)}{w - z_0} \, dw = \int_0^1 \frac{f(z_0)}{\rho \, e^{2\pi i t}} \, \rho \, 2\pi i \, e^{2\pi i t} \, dt$$

$$= 2\pi i \, f(z_0).$$

To estimate $\int_\gamma \frac{f(w) - f(z_0)}{w - z_0} \, dw$, we use the continuity of f at z_0.

Let $\varepsilon > 0$ be given. Then there is $\delta > 0$ such that $|f(z) - f(z_0)| < \varepsilon$ whenever $|z - z_0| < \delta$. Choose ρ so that $0 < \rho < \delta$. Then for any $w \in \operatorname{tr} \gamma$, we have that $|w - z_0| = \rho < \delta$ so that $|f(w) - f(z_0)| < \varepsilon$ and therefore $\left| \frac{f(w) - f(z_0)}{w - z_0} \right| = \frac{|f(w) - f(z_0)|}{\rho} < \frac{\varepsilon}{\rho}$. It follows that

$$\left| \int_\gamma \frac{f(w) - f(z_0)}{w - z_0} \, dw \right| \le \frac{\varepsilon}{\rho} L(\gamma)$$

$$= \frac{\varepsilon}{\rho} \, 2\pi\rho$$

$$= 2\pi\varepsilon.$$

Using this, we see that

$$\left| \int_\Gamma \frac{f(w)}{w - z_0} \, dw - 2\pi i f(z_0) \right| = \left| \int_\gamma \frac{f(w) - f(z_0)}{w - z_0} \, dw \right|$$

$$\le 2\pi\varepsilon.$$

This holds for any $\varepsilon > 0$. Hence, the left hand side (which, incidentally, does not depend on ρ) must be zero. $\qquad\square$

8.5 Taylor Series Expansion

The next theorem is yet another result of fundamental importance.

Theorem 8.6 (Taylor Series Expansion) *Suppose that f is analytic in the disc $D(z_0, R)$. Then, for any $z \in D(z_0, R)$, the function $f(z)$ has the power series expansion*

$$f(z) = \sum_{n=0}^\infty a_n (z - z_0)^n$$

where $a_n = \dfrac{1}{2\pi i} \displaystyle\int_\Gamma \frac{f(w)}{(w - z_0)^{n+1}} \, dw$ for any circle Γ encircling z_0 and lying in $D(z_0, R)$. The series converges absolutely for any $z \in D(z_0, R)$.

Proof.   Let $z \in D(z_0, R)$ be given. Let $r > 0$ satisfy $|z - z_0| < r < R$ and put $\gamma(t) = z_0 + re^{2\pi i t}$, for $0 \leq t \leq 1$. Notice, firstly, that $|z - z_0|/r < 1$ and, secondly, that z lies inside the circle γ.

The idea of the proof is to apply Cauchy's Integral Formula

$$f(z) = \frac{1}{2\pi i} \int_\gamma \frac{f(w)}{w - z}\, dw$$

and expand $\dfrac{1}{w - z}$ in powers of $(z - z_0)$.

The formula $1 - \alpha^n = (1 - \alpha)(1 + \alpha + \alpha^2 + \cdots + \alpha^{n-1})$ can be rewritten to give

$$\frac{1}{1 - \alpha} = 1 + \alpha + \alpha^2 + \cdots + \alpha^{n-1} + \alpha^n \frac{1}{1 - \alpha}.$$

Hence, we may write

$$\frac{1}{w - z} = \frac{1}{(w - z_0) - (z - z_0)} = \frac{1}{(w - z_0)} \left(\frac{1}{1 - \left(\dfrac{z - z_0}{w - z_0}\right)} \right)$$

$$= \frac{1}{(w - z_0)} \left\{ 1 + \left(\frac{z - z_0}{w - z_0}\right) + \left(\frac{z - z_0}{w - z_0}\right)^2 + \cdots \right.$$

$$\left. \cdots + \left(\frac{z - z_0}{w - z_0}\right)^{n-1} + \left(\frac{z - z_0}{w - z_0}\right)^n \frac{1}{1 - \left(\dfrac{z - z_0}{w - z_0}\right)} \right\}$$

$$= \frac{1}{(w - z_0)} + \frac{(z - z_0)}{(w - z_0)^2} + \cdots + \frac{(z - z_0)^{n-1}}{(w - z_0)^n} + \frac{(z - z_0)^n}{(w - z_0)^n (w - z)}.$$

Therefore, using Cauchy's Integral Formula, we have

$$f(z) = \frac{1}{2\pi i} \int_\gamma \frac{f(w)}{(w - z)}\, dw$$

$$= \frac{1}{2\pi i} \int_\gamma \frac{f(w)}{(w - z_0)}\, dw + \frac{1}{2\pi i} \int_\gamma \frac{f(w)}{(w - z_0)^2} (z - z_0)\, dw + \cdots$$

$$\cdots + \frac{1}{2\pi i} \int_\gamma \frac{f(w)}{(w - z_0)^n} (z - z_0)^{n-1}\, dw + \cdots$$

$$\cdots + \underbrace{\frac{1}{2\pi i} \int_\gamma \frac{f(w)(z - z_0)^n}{(w - z_0)^n (w - z)}\, dw}_{R_n}$$

$$= a_0 + a_1(z - z_0) + \cdots + a_{n-1}(z - z_0)^{n-1} + R_n,$$

where $a_k = \dfrac{1}{2\pi i} \displaystyle\int_\gamma \dfrac{f(w)}{(w-z_0)^{k+1}}\,dw$, for $k = 0, 1, 2, \ldots$ The value of a_k is independent of γ as long as it encircles the point z_0. This follows from the Deformation Lemma, lemma 8.1, with $\zeta = z_0$ and $g(w) = \dfrac{f(w)}{(w-z_0)^{k+1}}$.

We wish to show that $R_n \to 0$ as $n \to \infty$. To see this, we note that the continuity of f and the compactness of $\operatorname{tr}\gamma$ together imply that there is some $M > 0$ such that $|f(w)| \le M$ for any $w \in \operatorname{tr}\gamma$ (proposition 4.3). Also, $z \notin \operatorname{tr}\gamma$ and so there is $\delta > 0$ such that $|w - z| \ge \delta$ for any $w \in \operatorname{tr}\gamma$. (In fact, δ could be any positive real number less than $r - |z - z_0|$.) This means that $1/|w - z| \le 1/\delta$, whenever $w \in \operatorname{tr}\gamma$. Therefore

$$\left| \frac{f(w)}{(w-z_0)^n(w-z)} \right| \le \frac{M}{r^n \delta}$$

for any $w \in \operatorname{tr}\gamma$ (since $|w - z_0| = r$ if $w \in \operatorname{tr}\gamma$). Hence, we find

$$|R_n| = \left| \frac{1}{2\pi i} \int_\gamma \frac{f(w)(z-z_0)^n}{(w-z_0)^n(w-z)}\,dw \right|$$

$$\le \frac{1}{2\pi} |z - z_0|^n \frac{M}{r^n \delta} 2\pi r$$

$$= \frac{Mr}{\delta} \left(\frac{|z - z_0|}{r} \right)^n \to 0$$

as $n \to \infty$, since $\dfrac{|z - z_0|}{r} < 1$. Hence

$$f(z) = \sum_{n=0}^{\infty} a_n (z - z_0)^n,$$

as claimed.

We still have to establish the absolute convergence of this series. This follows from the convergence of the power series, but can also be shown directly. We estimate

$$|a_n(z-z_0)^n| = \left| \frac{1}{2\pi i} \int_\gamma \frac{f(w)(z-z_0)^n}{(w-z_0)^{n+1}}\,dw \right|$$

$$\le \frac{1}{2\pi} \frac{M |z - z_0|^n}{r^{n+1}} 2\pi r$$

$$= M \left(\frac{|z - z_0|}{r} \right)^n.$$

But $|z - z_0|/r < 1$ and so the series $\sum_{n=0}^{\infty} |a_n(z-z_0)^n|$ converges for any $z \in D(z_0, R)$, by the Comparison Test, as required. $\qquad\square$

8.6 Cauchy's Integral Formulae for Derivatives

As a direct consequence of this theorem, we shall show that any analytic function is differentiable to all orders and, moreover, that these derivatives are given by integral formulae. So we see that despite its innocuous formulation, analyticity is a very strong property.

Theorem 8.7 (Cauchy's Formula for derivatives)   *Let D be a domain and suppose that $f \in H(D)$. Then f has derivatives of all orders at every point in D. Furthermore, if $D(z_0, R) \subseteq D$, then $f^{(k)}(z_0)$, the kth-derivative of f at the point $z_0 \in D$, is given by the formula*

$$f^{(k)}(z_0) = \frac{k!}{2\pi i} \int_\Gamma \frac{f(w)}{(w - z_0)^{k+1}} \, dw$$

where Γ is any simple circle in $D(z_0, R)$ which encircles z_0 and $k \in \mathbb{N}$.

Proof.   For any given $z_0 \in D$, there is $R > 0$ such that $D(z_0, R) \subseteq D$. By theorem 8.6, f has a Taylor series (power series) expansion

$$f(z) = \sum_{n=0}^{\infty} a_n (z - z_0)^n,$$

valid for any $z \in D(z_0, R)$. The series converges absolutely for every z in the disc $D(z_0, R)$ and a_n is given by

$$a_n = \frac{1}{2\pi i} \int_\Gamma \frac{f(w)}{(w - z_0)^{n+1}} \, dw.$$

This means that inside the disc $D(z_0, R)$, the function f *is* a power series. Any power series has derivatives of all orders, and so this is true of f (in this disc). In particular, for any $k = 0, 1, 2, \ldots,$

$$f^{(k)}(z_0) = k! \, a_k.$$

Substituting the integral formula for a_k, as given by theorem 8.6, we obtain the required formula for the kth-derivative of f.   □

Remark 8.2   This should be contrasted with the real case where there is no such corresponding result. In fact, there are real functions of a real variable which are infinitely differentiable but whose Taylor series has zero radius of convergence. Indeed, the function

$$f(x) = \begin{cases} 0, & x \le 0 \\ e^{-\frac{1}{x^2}}, & x > 0 \end{cases}$$

provides just such an example.

The Taylor Series Expansion Theorem says that any function analytic in any disc is expressible as a power series which is absolutely convergent in that disc. By definition, a domain, D say, is an open set and therefore any point in D is inside a disc also lying in D. Then any $f \in H(D)$ is expressible as an absolutely convergent power series in any such disc. We can say that *locally* f is a power series.

Of course, by considering different points in D we are led to different discs and therefore to different power series.

We can think of a domain as a (in general, infinite) collection of (often overlapping) discs. In the same way, we can think of an analytic function as a whole collection of power series, each absolutely convergent in some disc in the domain of analyticity of the function.

It is important to appreciate that it is part of the Taylor Expansion Theorem that the power series expression for the analytic function f (about the point z_0) is absolutely convergent in any disc $D(z_0, R)$ centred on z_0 and lying in the domain D. The coefficients of this power series are the derivatives of the function f (together with extra $k!$ factors) and, in turn, are given by integral formulae. In particular, it is all part of the theorem that the series $\sum_{k=0}^{\infty} f^{(k)}(z_0)(z - z_0)^k/k!$ converges absolutely in the disc $D(z_0, R)$.

Finally, we note that these integral formulae for f and its derivatives at the point z_0 only involve the values of f on the contour Γ. This means that the values of an analytic function (and all its derivatives) are determined by the values of the function possibly quite far away. For example, if f is entire, then we can take Γ to be a circle of arbitrarily large radius about any given point z_0. The value of f (and each of its derivatives) at this point z_0 is determined by the values of f on this giant circle. In particular, we cannot mess about with the values of f in a small region without spoiling analyticity. To put this another way, if two analytic functions happen to be equal on a circle (and are analytic in some disc which contains this circle), then they necessarily agree everywhere inside the circle (because their values inside are given by integrals around the circle). We will pursue this phenomenon later (see the Identity Theorem, theorem 8.12).

Example 8.4 We know that the function $w \mapsto \operatorname{Log} w$ is analytic in the cut-plane $\mathbb{C} \setminus \mathbb{R}_{\leq 0}$. Setting $w = 1 + z$, we find that the function $f :$ $z \mapsto \operatorname{Log}(1 + z)$ is analytic in the cut-plane $\mathbb{C} \setminus \{ z : z \text{ real, and } z \leq -1 \}$. In particular, f is analytic in the disc $D(0, 1)$ and so has a Taylor series expansion about $z = 0$ which we are assured, by the theorem, converges absolutely for every z in this unit disc. The derivative of f is $\dfrac{1}{1+z}$ and so we may compute all the necessary derivatives to deduce that

$$\mathrm{Log}(1+z) = z - \frac{z^2}{2} + \frac{z^3}{3} - \frac{z^4}{4} + \cdots + (-1)^{n+1}\frac{z^n}{n} + \cdots.$$

An application of the ratio test confirms that this series converges absolutely for all z with $|z| < 1$, as it should.

An alternative derivation of this expansion of $\mathrm{Log}(1+z)$ is to note that the power series converges absolutely for $|z| < 1$ and has derivative $\dfrac{1}{1+z}$ for $|z| < 1$. This is also true of $\mathrm{Log}(1+z)$, so their difference is constant on the open unit disc. However, both functions take the value 0 when $z = 0$ and so they are equal throughout the disc $D(0,1)$.

Corollary 8.1 *For $f \in H(D)$, the real functions $u(x,y) = \mathrm{Re}\, f(x+iy)$ and $v(x,y) = \mathrm{Im}\, f(x+iy)$ have partial derivatives of all orders at any (x,y) such that $x + iy \in D$. Moreover, these partial derivatives may be taken in any order (for example, $u_{xyx} = u_{xxy} = u_{yxx}$).*

Proof. We have seen that f is infinitely differentiable in D as a complex function of a complex variable. The complex derivative can be taken, in particular, in the "real direction" or in the "imaginary direction". That is,

$$f' = \partial_x(u+iv) = \frac{1}{i}\,\partial_y(u+iv),$$

which gives

$$f' = u_x + iv_x = -iu_y + v_y$$

and we get the Cauchy-Riemann equations as before. Denoting complex differentiation by D_z, we have

$$D_z = \partial_x = -i\partial_y$$

where D_z is to be applied to $f(z)$ and the partial derivatives are applied to the function $u(x,y) + iv(x,y)$. Any occurrence of D_z can be replaced by either ∂_x or $-i\partial_y$, and vice versa. It follows, by induction, that all partial derivatives of u and v exist and that

$$f^{(k)}(z) = (D_z^k f)(z) = L_k \ldots L_2\, L_1\big(u(x,y) + iv(x,y)\big)$$

where L_j, $j = 1, \ldots, k$, stands arbitrarily for either ∂_x or $-i\partial_y$. If we choose any m of these terms to be equal to $-i\partial_y$, then we obtain the equality

$$L_k \ldots L_2 L_1 \big(u(x,y) + iv(x,y)\big) = (D_z^k f)(z)$$
$$= (-i)^m \partial_x^{k-m} \partial_y^m \big(u(x,y) + iv(x,y)\big).$$

Equating real and imaginary parts shows that the order in which the partial derivatives are taken is immaterial. For example (with $k = 9$),

$$\partial_x(-i\partial_y)\partial_x^2(-i\partial_y)\partial_x^4\big(u(x,y) + iv(x,y)\big) = (-i)^2 \partial_x^7 \partial_y^2 \big(u(x,y) + iv(x,y)\big)$$

so that

$$\partial_x \partial_y \partial_x^2 \partial_y \partial_x^4 u(x,y) = \partial_x^7 \partial_y^2 u(x,y)$$

and

$$\partial_x \partial_y \partial_x^2 \partial_y \partial_x^4 v(x,y) = \partial_x^7 \partial_y^2 v(x,y) \,. \qquad \square$$

What's going on? Under the initial assumption that f is analytic in a star-domain D, it follows that f has a primitive there and so has zero integral around any closed contour in D. This leads to the Deformation Lemma, which in turn gives Cauchy's Integral Formula. From this, there follows the power series (Taylor series) expansion in discs. In particular, it follows that if a function is analytic in some domain (star-like or non star-like, with or without holes), then in fact, it is infinitely-differentiable there. (The point here of course is that as far as analyticity is concerned, one can restrict attention to discs.)

8.7 Morera's Theorem

Next, we consider a converse to Cauchy's Theorem.

Theorem 8.8 (Morera's Theorem) *Suppose $f : D \to \mathbb{C}$ is continuous on the domain D and that $\int_{\partial T} f = 0$ for every triangle T wholly contained in D. Then $f \in H(D)$.*

Proof. Let $D(z_0, R)$ be any disc in D. Then, by hypothesis, $\int_{\partial T} f = 0$ for every triangle T wholly contained in $D(z_0, R)$. But the disc $D(z_0, R)$ is a star-domain and so we deduce, just as in the proof of theorem 8.2, that f has a primitive there, F, say. Hence $f = F'$ on $D(z_0, R)$. But the analyticity of F implies, in particular, that F is twice differentiable in $D(z_0, R)$. That is to say, F' is analytic in $D(z_0, R)$. However, $F' = f$ in $D(z_0, R)$ and so f is analytic in $D(z_0, R)$. We conclude that $f \in H(D)$. $\square$

8.8　Cauchy's Inequality and Liouville's Theorem

The following result provides a bound on the derivatives of an analytic function in terms of a bound on the function itself.

Theorem 8.9 (Cauchy's Inequality) *Suppose that f is analytic in the disc $D(z_0, R)$ and that $|f(z)| \leq M$ for all $z \in D(z_0, R)$. Then*

$$|f^{(k)}(z_0)| \leq \frac{M\,k!}{R^k},$$

for any $k = 0, 1, 2, \ldots$.

Proof.　Let $0 < r < R$ and put $\gamma(t) = z_0 + re^{2\pi it}$, $0 \leq t \leq 1$. By Cauchy's Integral Formula

$$f^{(k)}(z_0) = \frac{k!}{2\pi i} \int_\gamma \frac{f(w)}{(w - z_0)^{k+1}}\, dw.$$

However, for any $w \in \operatorname{tr} \gamma$, we have that $|f(w)| \leq M$ and $|w - z_0| = r$ and so

$$\left| \frac{f(w)}{(w - z_0)^{k+1}} \right| \leq \frac{M}{r^{k+1}}.$$

Using this, we estimate the integral to get

$$\begin{aligned}
|f^{(k)}(z_0)| &\leq \frac{k!}{2\pi} \frac{M}{r^{k+1}} 2\pi r \\
&= \frac{k!\,M}{r^k}.
\end{aligned}$$

This holds for any $0 < r < R$, and the left hand side does not depend on r. Taking the limit $r \to R$ gives

$$|f^{(k)}(z_0)| \leq \frac{k!\,M}{R^k}$$

and the proof is complete.　　　　　　　　　　　　　　　　　□

Remark 8.3　There is no analogue of this for functions of a real variable. For example, let $f(x) = \sin(\lambda x)$. Then $|f(x)| \leq 1$ for all $x \in \mathbb{R}$ and every λ, but $f'(0) = \lambda$, which can be chosen as large as we wish.

Theorem 8.10 (Liouville's Theorem) *If f is entire and bounded, then f is constant. In other words, no entire function can be bounded unless it is constant.*

Proof. Suppose that f is entire and that $|f(z)| \le M$ for every $z \in \mathbb{C}$. Then f is analytic in $\mathbb{C}$ and so has a Taylor series expansion about $z_0 = 0$ which is valid for z in *any* disc $D(0, R)$, i.e., for all z,

$$f(z) = \sum_{n=0}^{\infty} a_n z^n, \quad \text{with } a_n = \frac{f^{(n)}(0)}{n!}.$$

Applying Cauchy's Inequality, theorem 8.9, to f in the disc $D(0, R)$, we find that

$$|f^{(n)}(0)| \le \frac{n!\, M}{R^n},$$

for any $n = 0, 1, 2, \ldots$. This holds for any $R > 0$. Fixing $n \ge 1$ and letting $R \to \infty$, we conclude that $f^{(n)}(0) = 0$. It follows that $f(z) = a_0$, that is, the function f is constant. $\square$

Example 8.5 Suppose that f is entire and satisfies $|f(z)| \le 1 + |z|^m$ for all $z \in \mathbb{C}$. It follows that $|f(z)| \le 1 + R^m$ for all $z \in D(0, R)$ and so Cauchy's inequality implies that

$$|f^{(n)}(0)| \le \frac{n!\,(1 + R^m)}{R^n} \to 0$$

as $R \to \infty$, for any $n > m$. So if $f(z) = \sum_{n=0}^{\infty} a_n z^n$ is the Taylor series expansion of f about $z_0 = 0$, then all the coefficients a_n with $n > m$ vanish. In other words, $f(z)$ is a polynomial of degree at most m.

We can use Liouville's Theorem to give a fairly painless proof of the Fundamental Theorem of Algebra.

Theorem 8.11 (Fundamental Theorem of Algebra) *Every non-constant complex polynomial p has a zero, that is, there is $z_0 \in \mathbb{C}$ such that $p(z_0) = 0$.*

Proof. We may write p as $p(z) = a_n z^n + a_{n-1} z^{n-1} + \cdots + a_1 z + a_0$, where $n \ge 1$ and $a_n \neq 0$. To show that p has a zero, we suppose the contrary and obtain a contradiction. Thus p is entire and, assuming it is never zero, $1/p$

is also entire. However

$$\frac{1}{p(z)} = \frac{1}{a_n z^n + a_{n-1} z^{n-1} + \cdots + a_1 z + a_0}$$

$$= \frac{1}{z^n \left(a_n + \dfrac{a_{n-1}}{z} + \cdots + \dfrac{a_1}{z^{n-1}} + \dfrac{a_0}{z^n} \right)}.$$

As $|z| \to \infty$, $\left(a_n + \dfrac{a_{n-1}}{z} + \cdots + \dfrac{a_1}{z^{n-1}} + \dfrac{a_0}{z^n} \right) \to a_n$ and so $\dfrac{1}{p(z)} \to 0$. In

particular, it follows that there is R such that $\left| \dfrac{1}{p(z)} \right| < 1$, whenever $|z| > R$.

On the other hand, if p is never zero, then $1/p$ is analytic and so is certainly continuous. In particular, $1/p$ is bounded on the closed (and so compact) disc $\overline{D(0,R)}$, that is, there is $M > 0$ such that $\left| \dfrac{1}{p(z)} \right| \leq M$, whenever $|z| \leq R$.

Combining these remarks, we may say that $\dfrac{1}{p(z)}$ is entire and obeys $\left| \dfrac{1}{p(z)} \right| \leq M + 1$ for any $z \in \mathbb{C}$. Hence by Liouville's Theorem, $1/p$ is constant, say $1/p = \alpha$. But then $p = 1/\alpha$ is also constant, a contradiction. We conclude that p does, indeed, possess a zero, that is, $p(z_0) = 0$ for some $z_0 \in \mathbb{C}$. $\qquad\square$

Corollary 8.2   *Let* $p(z) = a_n z^n + \cdots + a_1 z + a_0$, *with* $a_n \neq 0$, *be a polynomial of degree* n. *Then there exist* $\zeta_1, \ldots, \zeta_n \in \mathbb{C}$ *and* $\alpha \in \mathbb{C}$ *such that*

$$p(z) = \alpha(z - \zeta_1) \ldots (z - \zeta_n)$$

for all $z \in \mathbb{C}$. *(N.B. The* ζ_js *need not be all different.) In other words, any polynomial of degree* n *has exactly* n *zeros—counted according to their multiplicity.*

Proof.   We shall prove this by induction. For each $n \in \mathbb{N}$, let $Q(n)$ be the statement that any polynomial of degree n can be written in the stated form. Any polynomial of degree 1 has the form $p(z) = az + b$ with $a \neq 0$. Clearly, such p can alternatively be written as $p(z) = a(z - c)$, where $c = -b/a$. Hence $Q(1)$ is true.

Next we show that the truth of $Q(n)$ implies that of $Q(n+1)$. So suppose that $Q(n)$ is true. Let p be any polynomial of degree $n+1$,

$$p(z) = a_{n+1} z^{n+1} + a_n z^n + \cdots + a_0.$$

By the theorem, p has a zero, z_0, say. Then $p(z_0) = 0$ and so

$$
\begin{aligned}
p(z) &= p(z) - p(z_0) \\
&= a_{n+1}z^{n+1} - a_{n+1}z_0^{n+1} + a_n z^n - a_n z_0^n + \cdots + a_1 z - a_1 z_0 + a_0 - a_0 \\
&= a_{n+1}(z^{n+1} - z_0^{n+1}) + a_n(z^n - z_0^n) + \cdots + a_1(z - z_0) \\
&= (z - z_0)\{a_{n+1}(z^n + z^{n-1}z_0 + \cdots + z_0^n) \\
&\qquad + a_n(z^{n-1} + z^{n-2}z_0 + \cdots + z_0^{n-1}) + \cdots + a_2(z + z_0) + a_1\} \\
&= (z - z_0)q(z)
\end{aligned}
$$

where $q(z)$ is a polynomial of degree n. By induction hypothesis, namely, that $Q(n)$ is true, we can write $q(z)$ as

$$ q(z) = \beta(z - \zeta_1)\ldots(z - \zeta_n) $$

for some β and $\zeta_1, \ldots, \zeta_n \in \mathbb{C}$. It follows that $Q(n + 1)$ is true. Hence, by induction, $Q(n)$ is true for all $n \in \mathbb{N}$. $\qquad\square$

A further direct consequence of Liouville's Theorem is the interesting observation that the range of any non-constant entire function permeates the complex plane, $\mathbb{C}$.

Proposition 8.1 *Suppose that f is entire and not constant. Then for any $w \in \mathbb{C}$ and any $\varepsilon > 0$ there is some $\zeta \in \mathbb{C}$ such that $f(\zeta) \in D(w, \varepsilon)$. In other words, f assumes values arbitrarily close to any complex number.*

Proof. Suppose that $w \in \mathbb{C}$ and $\varepsilon > 0$ are given. To say that there is no ζ with $f(\zeta)$ in the disc $D(w, \varepsilon)$ is to say that $|f(z) - w| \geq \varepsilon$ for all $z \in \mathbb{C}$. In particular, $f(z) - w \neq 0$ and so $g = 1/(f - w)$ is entire. But then g obeys $|g(z)| \leq 1/\varepsilon$ for all $z \in \mathbb{C}$ and so is constant, by Liouville's Theorem. It follows that $f - w$ and hence also f is constant. The result follows. $\qquad\square$

8.9 Identity Theorem

We have seen, in theorem 4.5, that if power series agree at a sequence of points converging to the centre of a common disc of convergence then they are, in fact, identical. In view of the results above, to the effect that analytic functions can be thought of as families of power series based on overlapping discs of convergence, it will come as no surprise that this theorem has an extension to functions analytic in a domain.

Recall that the complex number ζ is a limit point of a set S if and only if there is a sequence $(z_n)_{n\in\mathbb{N}}$ in S with $z_n \neq \zeta$, such that $z_n \to \zeta$, as $n \to \infty$. That is, ζ is the limit of some sequence from $S \setminus \{\zeta\}$.

Theorem 8.12 (Identity Theorem) *Suppose that D is a domain and that $f, g \in H(D)$. Suppose that there is some set $S \subseteq D$ such that S has a limit point which belongs to D and such that $f = g$ on S. Then $f = g$ on the domain D.*

Proof. Set $h = f - g$. Then $h \in H(D)$ and $h(z) = 0$ for all $z \in S$. We must show that $h(z) = 0$ for all $z \in D$. By hypothesis, there is some point $\zeta_0 \in D$ which is a limit point of S. Hence there is a sequence $(z_n)_{n\in\mathbb{N}}$ in S with $z_n \neq \zeta_0$, $n \in \mathbb{N}$, and such that $z_n \to \zeta_0$, as $n \to \infty$. Each $h(z_n) = 0$ and, by continuity, $h(\zeta_0) = 0$.

Let Z denote the zeros of h, $Z = \{\, z \in D : h(z) = 0 \,\}$. Clearly, $S \subseteq Z$ and we have just shown that $\zeta_0 \in Z$ and, moreover, ζ_0 is a limit point of Z. We will appeal to the connectedness of D. To this end, let

$$A = \{\, z \in D : z \text{ is a limit point of } Z \,\}$$

and

$$B = D \setminus A.$$

As noted above, ζ_0 is a limit point of Z and so belongs to A. In particular, $A \neq \varnothing$. It is also clear (by continuity) that h vanishes on A. We shall show that A is open. Indeed, suppose that $\zeta \in A$. Since $\zeta \in D$, there is $r > 0$ such that $D(\zeta, r) \subseteq D$. By theorem 8.6, h has a Taylor series expansion, absolutely convergent in $D(\zeta, r)$. Since $\zeta \in A$, it follows from theorem 4.5 that h vanishes in the whole disc $D(\zeta, r)$. But then each point w of the disc $D(\zeta, r)$ is clearly a limit point of Z (for example, take a sequence moving radially outwards towards w) which implies that $D(\zeta, r) \subseteq A$. It follows that A is open.

We wish to show that $B = \varnothing$, so that $A = D$. Suppose that $B \neq \varnothing$. We shall show that B is open. Let $w \in B$. Then $w \in D$ and so there is $R > 0$ such that $D(w, R) \subseteq D$. However, w is a member of B and so is not a limit point of zeros of h. This means that there is some $0 < \rho < R$ such that $h(z) \neq 0$ for all $z \in D$ with $0 < |w - z| < \rho$, i.e., h cannot vanish in some punctured disc around w. (Otherwise, h would vanish at some point, distinct from w, in every disc $D(w, r)$, $0 < r < R$, which would force w to be a limit point of zeros of h.) In particular, $D(w, \rho) \subseteq B$ and it follows that B is open.

Now, we appeal to the connectedness of D. We have seen that A and B are both open, they are disjoint and $D = A \cup B$. Hence one or the other must be empty. We know that $A \neq \varnothing$ and so we deduce that $B = \varnothing$. Hence $A = D$ and so h vanishes on D (because it vanishes on A).

Alternative Proof. This version uses directly the pathwise connectedness of D. As above, we wish to show that $h = f - g$ vanishes on the domain D. As before, let Z denote the set of zeros of h and let A denote the set of limit points of Z in D. By hypothesis, A is not empty, so let $z_0 \in A$ and let w be any point of D. We shall show that $h(w) = 0$. To see this, first note that $h(z) = 0$ for any $z \in A$. This is simply because h is continuous and so $z_n \to z$ means that $h(z_n) \to h(z)$ and therefore $h(z) = 0$ if each $h(z_n) = 0$.

Next, we note that since D is pathwise connected, there is some path $\varphi : [a, b] \to \mathbb{C}$ joining z_0 to w in D. Let $g : [a, b] \to \{0, 1\}$ be given by

$$g(t) = \begin{cases} 0, & \text{if } \varphi(t) \in A \\ 1, & \text{if } \varphi(t) \notin A. \end{cases}$$

Evidently $g(a) = 0$ because $\varphi(a) = z_0 \in A$. We shall show that the real-valued function g is continuous on the interval $[a, b]$. Let $t_0 \in [a, b]$ be given and let $w_0 = \varphi(t_0)$. There are two possibilities.

Case (i): $w_0 \in A$.

This means that $g(t_0) = 0$. Since $w_0 \in D$ there is some $r > 0$ such that $D(w_0, r) \subseteq D$. Now, $h(z)$ is analytic in the disc $D(w_0, r)$ and so has a Taylor series expansion about w_0 valid for all $z \in D(w_0, r)$. But $w_0 \in A$, so by the Identity Theorem for power series, we conclude that h vanishes throughout $D(w_0, r)$ and therefore every point of $D(w_0, r)$ is a limit point of Z.

Now, φ is continuous at t_0, and therefore there is some $\delta > 0$ such that $\varphi(t) \in D(w_0, r)$ whenever $|t - t_0| < \delta$ and $t \in [a, b]$. That is, $\varphi(t) \in A$ and so $g(t) = 0$ for all such t. It follows that g is continuous at t_0. (g is constant in a neighbourhood of t_0.)

Case (ii): $w_0 \notin A$.

This means that w_0 is not a limit point of zeros of h (but this does not preclude the possibility that $h(w_0) = 0$). It follows that there is some $\rho > 0$ such that h is never zero in the punctured disc $D'(w_0, \rho)$ (otherwise w_0 *would* be a limit point of zeros of h).

Furthermore, by construction, $g(t_0) = 1$. Once again the continuity of φ at t_0 means that there is $\delta' > 0$ such that $\varphi(s) \in D(w_0, \rho)$ whenever

$|s - t_0| < \delta'$ and $s \in [a, b]$. However, for any such s, either $\varphi(s) = w_0$, in which case $g(s) = 1$ (because $\varphi(s) = w_0 \notin A$) or $\varphi(s) \in D'(w_0, \rho)$ and so $h(\varphi(s)) \neq 0$ and therefore $g(s) = 1$ because $\varphi(s)$ is certainly not a limit point of zeros of h. It follows that g is continuous at t_0.

We see then, that g is continuous at each $t_0 \in [a, b]$. Now, g only assumes at most two values, either 0 or 1. By the Intermediate Value Theorem, it follows that g is constant. But $g(a) = 0$ and so $g(t) = 0$ for all $t \in [a, b]$. Therefore $g(b) = 0$ and so $w = \varphi(b) \in A$ and, in particular, $h(w) = 0$. This completes the proof. □

Remark 8.4　In practice, the set S is usually a line segment, or a disc or part of a disc or some similarly simple geometric region.

This result expresses a remarkable rigidity property enjoyed by analytic functions. For example, suppose that f is entire. Suppose that g is also entire and that $g(z) = f(z)$ for z in, say, the very small and very remote line segment $S = [10^{66}, 10^{66} + 10^{-999}]$. Nonetheless, by the theorem, it follows that $g(z) = f(z)$ throughout the whole complex plane.

Corollary 8.3　*The complex functions* $\exp z$, $\cos z$ *and* $\sin z$ *are the only entire functions which agree with their real counterparts* $\exp x$, $\cos x$ *and* $\sin x$, *respectively, on the real axis. That is, if f is entire and $f(x) = \exp x$ for all $x \in \mathbb{R}$, then $f(z) = \exp z$ for all $z \in \mathbb{C}$, and similarly for $\cos z$ and $\sin z$.*

Proof.　Take $S = \mathbb{R}$ and apply the theorem. □

Remark 8.5　Suppose that D is a domain and suppose that $f \in H(D)$ is not identically zero on D. Let S denote the (possibly empty) set of zeros of f;

$$S = \{\, \text{zeros of } f \,\} = \{\, z \in D : f(z) = 0 \,\}.$$

Then S can have no limit points *in D*—otherwise, by the Identity Theorem, f would be zero throughout D. In other words, the zeros of f are isolated: if $z_0 \in D$ is such that $f(z_0) = 0$ then there is $r > 0$ such that f has no other zeros in the disc $D(z_0, r)$.

Definition 8.2　Suppose that f is analytic in a domain D. The point $z_0 \in D$ is said to be a zero of f of order m (with $m \geq 1$) if $f(z_0) = 0$ and the Taylor series expansion of f about z_0 has the form $f(z) = \sum_{k=m}^{\infty} a_k(z - z_0)^k$ where $a_m \neq 0$.

This is equivalent to demanding that the derivatives $f(z_0)$, $f'(z_0)$, ..., $f^{(m-1)}(z_0)$ all vanish, but $f^{(m)}(z_0) \neq 0$. Note also, that according to the discussion above, the Identity Theorem implies that either f vanishes throughout D or every zero of f is isolated and so has some (finite) order. (In the latter case, the Taylor series cannot be the zero series and so must have at least one non-vanishing coefficient.)

As the next example shows, the set of zeros may well have a limit point not belonging to the domain.

Example 8.6 Let $D = \mathbb{C} \setminus \{0\}$, the punctured plane and let $f(z) = \sin(1/z)$ for $z \in D$. Now, $\sin(1/z) = 0$ whenever $1/z = k\pi$ for some $k \in \mathbb{Z}$, i.e., when $z = 1/(k\pi)$, $k \in \mathbb{Z} \setminus \{0\}$.

Let $S = \{ z : z = 1/(k\pi), \ k \in \mathbb{Z} \setminus \{0\} \}$. We see that $S \subseteq D$, f vanishes on S and that S has (the single) limit point 0. However, f does not vanish on the whole of D. This does not contradict the Identity Theorem because the limit point 0 does not belong to the domain D.

Recall that a set is said to be countable if it has a finite number of elements (or is empty) or if its elements can be listed as a sequence (that is, can be labelled by $\mathbb{N}$). For example, the sets $\mathbb{N}$, $\mathbb{Z}$ and $\mathbb{Q}$ are countable, but one can show that the sets $(0,1)$ and $\mathbb{R}$ are not.

Theorem 8.13 *Suppose that D is a domain and that $f \in H(D)$. Then either f vanishes throughout D or the set of zeros of f is countable.*

Proof. Suppose that f is not identically zero on D and let $Z = \{ z \in D : f(z) = 0 \}$ denote the set of zeros of f in D. Let $K \subseteq D$ be compact. We show that $K \cap Z$ is either empty or has only a finite number of elements. Indeed, suppose that $K \cap Z$ is an infinite set. Let $w_1 \in K \cap Z$. Now let $w_2 \in (K \cap Z) \setminus \{ w_1 \}$ and let $w_3 \in (K \cap Z) \setminus \{ w_1, w_2 \}$. Continuing in this way, we construct a sequence (w_n) in $K \cap Z$ such that $w_n \neq w_m$ for any $n \neq m$. (This construction works because $(K \cap Z) \setminus \{ w_1, w_2, \ldots, w_n \}$ is not empty.) Now, K is compact and so (w_n) has a convergent subsequence, $w_{n_k} \to \zeta$, say, with $\zeta \in K$. But each w_{n_k} is a zero of f and so ζ is a limit point of zeros in D. By the Identity Theorem, this is impossible and so we conclude that $K \cap Z$ cannot contain infinitely-many elements.

To complete the proof, we observe that by proposition 3.6, the set D has a compact exhaustion, $D = \bigcup_{n=1}^{\infty} K_n$ where each K_n is compact. But then $Z = \bigcup_{n=1}^{\infty} (K_n \cap Z)$ which is the union of a sequence of finite (or empty) sets and so is countable. $\qquad \square$

8.10   Preservation of Angles

Consider a path γ and a point z_0 on the path. Let us say that γ makes an angle $\widehat{\theta}$ with respect to the positive real direction at the point $z_0 = \gamma(t_0)$ if $\gamma(t) \neq z_0$ for sufficiently small $t - t_0 > 0$ and if

$$\frac{\gamma(t) - z_0}{|\gamma(t) - z_0|} \to e^{i\theta}$$

as $t \downarrow t_0$. The angle between two paths γ_2 and γ_1, each passing through z_0 is then $\widehat{\theta}_2 - \widehat{\theta}_1$, where $\widehat{\theta}_j$ is the angle γ_j makes with the positive real direction at z_0, for $j = 1, 2$.

Now suppose that f is analytic in some disc $D(z_0, r)$ and is not constant there. Then we know that f has a Taylor series expansion

$$f(z) = \sum_{n=0}^{\infty} a_n (z - z_0)^n.$$

By hypothesis, f is not constant and so there must be some non-zero a_n with $n \geq 1$. Suppose that a_m is the first such non-zero coefficient so that the Taylor series for f has the form

$$f(z) = \sum_{n=0}^{\infty} a_n (z - z_0)^n = a_0 + \sum_{n=m}^{\infty} a_n (z - z_0)^n = f(z_0) + (z - z_0)^m g(z)$$

where $m \geq 1$ and $g(z_0) = a_m \neq 0$.

Let $\Gamma_j(t) = f(\gamma_j(t))$ for $j = 1, 2$. Since f is not constant, the Identity Theorem implies that $\Gamma_j(t) \neq f(z_0)$ for all sufficiently small $t - t_0 > 0$ (where $z_0 = \gamma(t_0)$). Hence

$$\frac{\Gamma_j(t) - f(z_0)}{|\Gamma_j(t) - f(z_0)|} = \frac{f(\gamma_j(t)) - f(z_0)}{|f(\gamma_j(t)) - f(z_0)|}$$

$$= \left(\frac{\gamma_j(t) - z_0}{|\gamma_j(t) - z_0|} \right)^m \frac{g(\gamma_j(t))}{|g(\gamma_j(t))|}$$

$$\to (e^{i\theta_j})^m \frac{g(z_0)}{|g(z_0)|}, \quad \text{as } t \downarrow t_0,$$

$$= e^{im\theta_j} e^{i \operatorname{Arg} g(z_0)}$$

and so Γ_j makes an angle $m\widehat{\theta}_j + \operatorname{Arg} g(z_0)$ with respect to the positive real direction at $f(z_0)$. But then this means that the angle between Γ_2 and Γ_1 at $f(z_0)$ is $m(\widehat{\theta}_2 - \widehat{\theta}_1)$.

In other words, if the derivatives $f^{(r)}(z_0) = 0$ for all $r = 1, \ldots, m-1$, but $f^{(m)}(z_0) \neq 0$, then the angle between paths intersecting at z_0 is multiplied by m under f. In particular, if $f'(z_0) \neq 0$, then f preserves the angle between intersecting paths. (Such maps are said to be conformal.)

Chapter 9

The Laurent Expansion

9.1 Laurent Expansion

In this chapter, we discuss a generalization of the Taylor series expansion. Rather than considering analyticity in some disc, we assume only analyticity in an annulus. This leads to a representation in terms of both positive and, possibly, negative powers of z, the Laurent expansion. The starting point is Cauchy's Integral Formula for the given function f. Here the circle of integration can be deformed, as in the Deformation Lemma, to produce, in fact, two contour integrals. Each of these is the integral of $f(w)/(w-z)$ around a certain circle. The idea is then to expand $1/(w-z)$ in powers of $z - z_0$, in one case, and in powers of $1/(z - z_0)$ in the other. Doing the integrals gives the coefficients and all that remains is to worry a bit about the convergence of the two series thus obtained. Now for the details.

Lemma 9.1 *Suppose that f is analytic in the annulus $A = A(z_0; R_1, R_2) = \{z : R_1 < |z - z_0| < R_2\}$. Let $z \in A$, and let $R_1 < r_1 < |z - z_0| < r_2 < R_2$. Then*

$$2\pi i f(z) = -\int_{C_1} \frac{f(w)}{w - z}\, dw + \int_{C_2} \frac{f(w)}{w - z}\, dw,$$

where C_1 is the circle $z_0 + r_1 e^{2\pi i t}$ and C_2 is the circle $z_0 + r_2 e^{2\pi i t}$, $0 \le t \le 1$.

Proof. We argue just as in the proof of the Deformation Lemma, lemma 8.1. Insert line segments between the two circular contours, as shown in Fig. 9.1. The idea is to put in sufficiently many cross-cuts so that each of the contours γ_j is so narrow that it is contained in the star-domain D_j which is itself inside the annulus A. Clearly, if $r_2 - R_1$ is small, there will need to be many such cross-cuts. We number the γ_js so that γ_1 goes around the point z, as indicated.

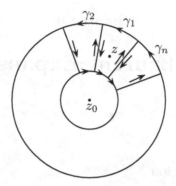

Fig. 9.1   Construct cross-cuts to give many narrow simple closed contours $\gamma_1, \gamma_2, \ldots, \gamma_n$.
The point z is encircled by γ_1.

Next, let Γ be a circle around z with sufficiently small radius that it is
encircled by γ_1, as shown in Fig. 9.2.

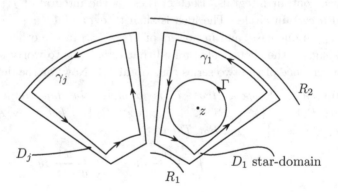

Fig. 9.2   Star-like domains D_j containing γ_j.

Using Cauchy's Integral Formula and arguing as in the proof of the
Deformation Lemma, lemma 8.1, we have

$$2\pi i f(z) = \int_\Gamma \frac{f(w)}{w - z}\, dw = \int_{\gamma_1} \frac{f(w)}{w - z}\, dw.$$

Furthermore, since $D_2, D_3, \ldots, D_n$ are star-like and $\dfrac{f(w)}{w-z}$ is analytic in each of these domains, it follows, by Cauchy's Theorem for a star-domain, theorem 8.4, that

$$\int_{\gamma_2} \frac{f(w)}{w-z}\, dw = \cdots = \int_{\gamma_n} \frac{f(w)}{w-z}\, dw = 0.$$

Piecing all these together, and using the fact that the contour integrals along the cross-cuts cancel out, we get

$$\int_{\gamma_1} \frac{f(w)}{w-z}\, dw + \int_{\gamma_2} \frac{f(w)}{w-z}\, dw + \cdots + \int_{\gamma_n} \frac{f(w)}{w-z}\, dw$$
$$= \int_{C_2} \frac{f(w)}{w-z}\, dw + \int_{C_1} \frac{f(w)}{w-z}\, dw.$$

We conclude that

$$2\pi i f(z) = \int_{C_2} \frac{f(w)}{w-z}\, dw - \int_{C_1} \frac{f(w)}{w-z}\, dw,$$

and the proof is complete. $\qquad\qquad\qquad\qquad\qquad\qquad\qquad\qquad\square$

As mentioned earlier, the next step is to expand $1/(w-z)$ into suitable powers. We use the formula

$$\frac{1}{1-\alpha} = 1 + \alpha + \alpha^2 + \cdots + \alpha^{k-1} + \frac{\alpha^k}{1-\alpha}$$

valid for any complex number $\alpha \neq 1$ and any $k \in \mathbb{N}$. Indeed, for any $n \in \mathbb{N}$, and $w \neq z_0$, $z \neq z_0$, we have

$$\frac{1}{w-z} = \frac{1}{(w-z_0)-(z-z_0)} = \frac{1}{(w-z_0)(1-(z-z_0)/(w-z_0))}$$
$$= \frac{1}{(w-z_0)}\left(\frac{z-z_0}{w-z_0} + \cdots + \left(\frac{z-z_0}{w-z_0}\right)^{n-1} + \frac{(z-z_0)^n}{(w-z_0)_{n-1}(w-z)}\right)$$
$$= \frac{1}{(w-z_0)} + \frac{(z-z_0)}{(w-z_0)^2} + \cdots + \frac{(z-z_0)^{n-1}}{(w-z_0)^n} + \frac{(z-z_0)^n}{(w-z_0)^n(w-z)}$$
$$\equiv S_n(z_0, z, w), \text{ say.}$$

Similarly, for any $m \in \mathbb{N}$ and $w \neq z_0$ and $z \neq z_0$, we have

$$-\frac{1}{w-z} = \frac{-1}{(w-z_0)-(z-z_0)} = \frac{1}{(z-z_0)-(w-z_0)}$$

$$= \frac{1}{(z-z_0)} + \frac{(w-z_0)}{(z-z_0)^2} + \cdots + \frac{(w-z_0)^{m-1}}{(z-z_0)^m} + \frac{(w-z_0)^m}{(z-z_0)^m(z-w)}$$

$$= S_m(z_0, w, z).$$

Notice that the first expression involves positive powers of $z - z_0$, whilst the second involves negative powers of $z - z_0$. Why should we be interested in both cases? The point is that $|w - z_0| < |z - z_0|$ whenever w belongs to the inner circle C_1, whereas the reverse inequality holds when w belongs to the outer circle C_2. This means that $\left|\frac{z-z_0}{w-z_0}\right| < 1$ in the first case so that $S_n(z_0, z, w)$ converges as $n \to \infty$. On the other hand, if w is on the outer circle C_2, then $\left|\frac{w-z_0}{z-z_0}\right| < 1$ and so the second expansion, $S_m(z_0, w, z)$, the one with negative powers of $(z - z_0)$, will converge as $m \to \infty$. Applying these considerations, together with lemma 9.1, leads to the Laurent expansion, as follows.

Theorem 9.1 (Laurent Expansion)   *Suppose that f is analytic in the annulus $A = A(z_0; R_1, R_2)$. Then, for any $z \in A$,*

$$f(z) = \sum_{n=0}^{\infty} a_n(z - z_0)^n + \sum_{n=1}^{\infty} b_n(z - z_0)^{-n}$$

with

$$a_n = \frac{1}{2\pi i} \int_C \frac{f(w)}{(w - z_0)^{n+1}} \, dw, \quad \textit{for } n = 0, 1, 2, \ldots,$$

and

$$b_n = \frac{1}{2\pi i} \int_C (w - z_0)^{n-1} f(w) \, dw = \frac{1}{2\pi i} \int_C \frac{f(w)}{(w - z_0)^{-n+1}} \, dw,$$

for $n = 1, 2, \ldots$, where C is any circle encircling z_0 such that $\operatorname{tr} C \subseteq A$.

Furthermore, both of these series are absolutely convergent in the given annulus A.

Proof.   Let $z \in A = A(z_0; R_1, R_2)$ be given and let r_1 and r_2 satisfy the inequalities $R_1 < r_1 < |z - z_0| < r_2 < R_2$. Let C_1 and C_2 be the circles $z_0 + r_1 e^{2\pi i t}$ and $z_0 + r_2 e^{2\pi i t}$, for $0 \le t \le 1$, respectively. Then using lemma

9.1, together with the preliminary discussion above, we see that $f(z)$ is given by

$$
\begin{aligned}
f(z) &= \frac{1}{2\pi i} \int_{C_2} \frac{f(w)}{w-z}\, dw - \frac{1}{2\pi i} \int_{C_1} \frac{f(w)}{w-z}\, dw \\
&= \frac{1}{2\pi i} \int_{C_2} f(w) S_n(z_0, z, w)\, dw + \frac{1}{2\pi i} \int_{C_1} f(w) S_m(z_0, w, z)\, dw \\
&= a_0 + a_1(z - z_0) + \cdots + a_{n-1}(z - z_0)^{n-1} + P_n \\
&\qquad + \frac{b_1}{(z - z_0)} + \cdots + \frac{b_m}{(z - z_0)^m} + Q_m
\end{aligned}
$$

where the as and bs are as stated in the theorem, and where the remainder terms P_n and Q_m are given by

$$
P_n = \frac{1}{2\pi i} \int_{C_2} \frac{(z - z_0)^n\, f(w)}{(w - z_0)^n\, (w - z)}\, dw
$$

and

$$
Q_m = \frac{1}{2\pi i} \int_{C_1} \frac{(w - z_0)^m\, f(w)}{(z - z_0)^m\, (z - w)}\, dw.
$$

We wish to show that $P_n \to 0$ as $n \to \infty$ and that $Q_m \to 0$ as $m \to \infty$. The traces of the circles C_1 and C_2 are compact sets in $\mathbb{C}$ and therefore there is some $M > 0$ such that $|f(w)| \leq M$ whenever $w \in \operatorname{tr} C_1 \cup \operatorname{tr} C_2$.

Furthermore, since $z \notin \operatorname{tr} C_1 \cup \operatorname{tr} C_2$, it follows that there is some $\delta > 0$ such that $D(z, \delta) \cap (\operatorname{tr} C_1 \cup \operatorname{tr} C_2) = \varnothing$ (because $z \in \mathbb{C} \setminus (\operatorname{tr} C_1 \cup \operatorname{tr} C_2)$, an open set). Hence $|w - z| \geq \delta$ for every $w \in \operatorname{tr} C_1 \cup \operatorname{tr} C_2$. (One could alternatively apply proposition 3.7 to reach this conclusion.)

We can use this to estimate $|P_n|$ and $|Q_m|$, as follows.

$$
|P_n| \leq \frac{1}{2\pi} \frac{|z - z_0|^n\, M}{r_2^n\, \delta}\, 2\pi r_2
$$

$$
\to 0,
$$

as $n \to \infty$, since $\dfrac{|z - z_0|}{r_2} < 1$. Similarly,

$$
|Q_m| \leq \frac{1}{2\pi} \frac{r_1^m}{|z - z_0|^m} \frac{M}{\delta}\, 2\pi r_1
$$

$$
\to 0,
$$

as $m \to \infty$, since $\dfrac{r_1}{|z - z_0|} < 1$. Taking the limit, say, $n \to \infty$, first, and then $m \to \infty$, it follows that

$$f(z) = \sum_{k=0}^{\infty} a_k(z - z_0)^k + \sum_{j=1}^{\infty} b_j(z - z_0)^{-j} \, ,$$

as claimed.

By an argument as in the proof of the Deformation Lemma, lemma 8.1, we see that the a_n and b_n integrals are independent of the particular circles C_1 and C_2, provided that they lie in A and encircle z_0.

To show that the two series are absolutely convergent in A, we simply estimate the general term in each case. We have

$$\begin{aligned}
|a_n(z - z_0)^n| &= \frac{1}{2\pi} \left| \int_{C_2} \frac{(z - z_0)^n}{(w - z_0)^{n+1}} f(w) \, dw \right| \\
&\leq \frac{1}{2\pi} \frac{|z - z_0|^n}{r_2^{n+1}} M \, 2\pi r_2 \\
&= M \left(\frac{|z - z_0|}{r_2} \right)^n ,
\end{aligned}$$

and

$$\begin{aligned}
|b_n(z - z_0)^{-n}| &= \frac{1}{2\pi} \left| \int_{C_1} \frac{(w - z_0)^{n-1}}{(z - z_0)^n} f(w) \, dw \right| \\
&\leq \frac{1}{2\pi} \frac{r_1^{n-1}}{|z - z_0|^n} M \, 2\pi r_1 \\
&= M \left(\frac{r_1}{|z - z_0|} \right)^n .
\end{aligned}$$

By the Comparison Test, both series converge absolutely. $\qquad\square$

Definition 9.1 The (double) series constructed above for the function f, analytic in A, is called the Laurent series expansion of f in the annulus $A(z_0; R_1, R_2)$. The series of negative powers, $\sum_{n=1}^{\infty} b_n(z - z_0)^{-n}$ is called the principal part of f.

Our next task is to establish the uniqueness of the Laurent expansion. After all, it is not obvious that it is not possible to change some (or all) of the a_ns and some (or all) of the b_ns without affecting f.

As a preliminary observation, we note that if $\gamma(t) = z_0 + re^{2\pi it}$, for $0 \le t \le 1$, is the circular contour centred at z_0 and with radius r, then

$$\frac{1}{2\pi i} \int_\gamma \frac{dz}{(z - z_0)^m} = \begin{cases} 0, & m \in \mathbb{Z}, \ m \ne 1 \\ 1, & m = 1. \end{cases}$$

This is a consequence of corollary 7.2, since $(z - z_0)^{-m}$ has a primitive in $\mathbb{C} \setminus \{z_0\}$ provided $m \ne 1$.

9.2 Uniqueness of the Laurent Expansion

Theorem 9.2 (Uniqueness of Laurent Expansion) *Suppose that f is analytic in the annulus $A = A(z_0; R_1, R_2)$ and that*

$$f(z) = \sum_{n=0}^{\infty} \alpha_n (z - z_0)^n + \sum_{n=1}^{\infty} \beta_n (z - z_0)^{-n}$$

where each of the two series converges absolutely in A. Then $\alpha_n = a_n$ and $\beta_n = b_n$ for all n, where the a_ns and b_ns are the Laurent series coefficients (as given by the integral formulae in theorem 9.1).

Proof. Let $\varepsilon > 0$ be given and set $R_n(z) = \sum_{k=n+1}^{\infty} \alpha_k (z - z_0)^k$ and set $T_n(z) = \sum_{k=n+1}^{\infty} \beta_k (z - z_0)^{-k}$, for $n \in \mathbb{N}$. The series for R_n converges absolutely in A, i.e., for z with $R_1 < |z - z_0| < R_2$. It follows that this series converges for all z with $|z - z_0| < R_2$. In other words, R_n is analytic in the disc $D(z_0, R_2)$, and, in particular, R_n is analytic in A.

However,

$$T_n(z) = f(z) - \sum_{k=0}^{n} \alpha_k (z - z_0)^k - R_n(z) - \sum_{k=1}^{n} \beta_k (z - z_0)^{-k}$$

and so T_n is analytic in the annulus A since this is true of the right hand side. In particular, if $R_1 < r < R_2$ and $\gamma(t) = z_0 + re^{2\pi it}$, $0 \le t \le 1$, then both R_n and T_n are continuous on tr γ.

Furthermore, $\sum_{k=0}^{\infty} |\alpha_k| r^k$ and $\sum_{k=1}^{\infty} |\beta_k| r^{-k}$ both converge and so there is $N \in \mathbb{N}$ such that

$$\sum_{k=n+1}^{\infty} |\alpha_k| r^k < \varepsilon \quad \text{and} \quad \sum_{k=n+1}^{\infty} |\beta_k| r^{-k} < \varepsilon$$

whenever $n > N$. It follows that, for any $z \in \operatorname{tr} \gamma$ (so that $|z - z_0| = r$),

$$|R_n(z)| \leq \sum_{k=n+1}^{\infty} |\alpha_k| r^k < \varepsilon$$

and that

$$|T_n(z)| \leq \sum_{k=n+1}^{\infty} |\beta_k| r^{-k} < \varepsilon$$

whenever $n > N$.

Finally, let $m \in \mathbb{Z}$ be given. Choose $n > |m|$. We have

$$f(z) = \sum_{k=0}^{n} \alpha_k (z - z_0)^k + \sum_{k=1}^{n} \beta_k (z - z_0)^{-k} + R_n(z) + T_n(z).$$

Hence

$$\frac{1}{2\pi i} \int_{\gamma} \frac{f(z)}{(z - z_0)^{m+1}} \, dz = \left\{ \begin{array}{l} \alpha_m, \text{ if } m \geq 0 \\ \beta_{|m|}, \text{ if } m < 0 \end{array} \right\} + \frac{1}{2\pi i} \int_{\gamma} \frac{R_n(z) + T_n(z)}{(z - z_0)^{m+1}} \, dz.$$

But the left hand side is just $\left\{ \begin{array}{l} a_m, \text{ if } m \geq 0 \\ b_{|m|}, \text{ if } m < 0 \end{array} \right\}$, according to their definitions.

Furthermore,

$$\left| \frac{1}{2\pi i} \int_{\gamma} \frac{R_n(z) + T_n(z)}{(z - z_0)^{m+1}} \, dz \right| \leq \frac{1}{2\pi} \frac{2\varepsilon}{r^{m+1}} 2\pi r, \quad \text{for all } n > N,$$

$$= \frac{2\varepsilon}{r^m}.$$

This is an estimate for the modulus of the difference between a_m and α_m or between $b_{|m|}$ and $\beta_{|m|}$, depending on whether $m \geq 0$ or not (and assuming that n is chosen greater than both N and $|m|$). Since this holds for any given $\varepsilon > 0$, we conclude that $\alpha_m = a_m$ for $m = 0, 1, 2, \ldots$ and $\beta_m = b_m$ for $m = 1, 2, \ldots$, as required. $\square$

Remark 9.1 Suppose that $r_1 < r < R < R_1$ and that f is analytic in the annulus $A(z_0; r_1, R_1)$. Then the uniqueness of the Laurent expansion implies that the coefficients in the Laurent expansion of f in the annulus $A(z_0; r, R)$ are precisely the same as those in the Laurent expansion of f in the annulus $A(z_0; r_1, R_1)$.

What's going on ?   A function f analytic in an annulus can be written as a sum of a series of positive powers and a series of negative powers. Both series converge absolutely in the annulus and the coefficients can be expressed in terms of the function f by means of certain contour integrals. This is the content of the Laurent Expansion Theorem. The absolute convergence of these series is part of the theorem. The uniqueness of the Laurent expansion means that whenever and however one manages to write f as a sum of absolutely convergent powers and inverse powers, then this has to be the Laurent expansion.

Examples 9.1

(1) For any $z \neq 0$, the power series definition of the sine function gives

$$\sin \frac{1}{z^2} = \frac{1}{z^2} - \frac{1}{3! \, z^6} + \frac{1}{5! \, z^{10}} - \frac{1}{7! \, z^{14}} + \cdots$$

which is *the* Laurent expansion about $z = 0$ of the function $\sin(1/z^2)$, valid for z in the punctured plane $\mathbb{C} \setminus \{0\}$.

(2) $f(z) = 1/(z - a)$ is analytic in the annulus $\{\, z : |z - a| > 0 \,\}$. The Laurent expansion of f about $z = a$, valid for $|z - a| > 0$, is simply

$$f(z) = \frac{1}{z - a}.$$

The function f is also analytic in the annulus $\{\, z : |z| > |a| \,\}$. We find

$$f(z) = \frac{1}{z(1 - a/z)} = \frac{1}{z}\left(1 + \frac{a}{z} + \frac{a^2}{z^2} + \cdots\right) = \frac{1}{z} + \frac{a}{z^2} + \frac{a^2}{z^3} + \cdots$$

is *the* Laurent expansion of f about $z = 0$, valid for $|z| > |a|$.
For z in $\{\, z : 0 < |z| < |a| \,\}$, we find that

$$f(z) = \frac{1}{z - a} = -\frac{1}{a(1 - z/a)} = -\frac{1}{a} - \frac{z}{a^2} - \frac{z^2}{a^3} - \cdots$$

is *the* Laurent expansion of f about $z = 0$, valid for z with $0 < |z| < |a|$ (where we suppose that $a \neq 0$). This is also valid for $z = 0$.
Note that these series converge absolutely for the given ranges of z.

(3) The function $f(z) = z(e^{1/z} - 1)$ is analytic in the punctured plane $\mathbb{C} \setminus \{0\}$. For any $z \neq 0$, we have

$$f(z) = z\left(\left(1 + \frac{1}{z} + \frac{1}{2! \, z^2} + \frac{1}{3! \, z^3} + \cdots\right) - 1\right)$$

$$= 1 + \frac{1}{2! \, z} + \frac{1}{3! \, z^2} + \frac{1}{4! \, z^3} + \cdots$$

which is the Laurent expansion of f about $z = 0$.

Proposition 9.1   *Suppose f is analytic and bounded in the punctured disc $D'(z_0, R)$. Then there is a function F analytic in the whole disc $D(z_0, R)$ such that $f = F$ on $D'(z_0, R)$. In other words, f can be extended so that it is defined at z_0 in such a way that the resulting function is analytic (in the whole disc).*

Proof.   The hypotheses imply that f has a Laurent series expansion in the annulus $A(z_0; r, R)$ for any $0 < r < R$. In particular, we can calculate the coefficients of the principal part of the Laurent expansion as

$$b_m = \frac{1}{2\pi i} \int_\gamma (z - z_0)^{m-1} f(z) \, dz$$

where $\gamma(t) = z_0 + \rho e^{2\pi i t}$ for $0 \leq t \leq 1$ and ρ can be chosen arbitrarily in the range $0 < \rho < R$.

By hypothesis, there is $M > 0$ such that $|f(z)| \leq M$ for all z in $D'(z_0, R)$. Hence, for any $m \geq 1$,

$$|b_m| \leq \frac{1}{2\pi} \rho^{m-1} M \, 2\pi\rho = M \rho^m.$$

This holds for any $0 < \rho < R$ and so it follows that $b_m = 0$ for all $m \geq 0$. The principal part of the Laurent expansion of f vanishes and we have

$$f(z) = \sum_{n=0}^{\infty} a_n (z - z_0)^n, \text{ for } z \in D'(z_0, R).$$

This series converges absolutely for all $0 < |z - z_0| < R$ and so certainly for all $|z - z_0| < R$. Hence, we may define F on the disc $D(z_0, R)$ by the power series

$$F(z) = \sum_{n=0}^{\infty} a_n (z - z_0)^n.$$

The absolute convergence implies that F is analytic in $D(z_0, R)$. Clearly $F = f$ on the punctured disc $D'(z_0, R)$ (and $F(z_0) = a_0$).   □

What's going on?   If f is analytic in a punctured disc in which it is also bounded, then, by suitable adjustment of f at the centre of this punctured disc, one finds that f is really just the restriction of a function analytic in the whole disc to the punctured disc. This is because one can estimate the coefficients occurring in the principal part of Laurent expansion and show that they must vanish. In other words, f is just a sum of positive powers, that is, it is a power series and we know that power series are analytic. We will need this result later.

Chapter 10

Singularities and Meromorphic Functions

10.1 Isolated Singularities

We can think of the principal part of the Laurent expansion of a function f analytic in a punctured disc as encapsulating how badly behaved f is near to the centre. There are essentially only three situations; all coefficients in the principal part vanish, only a finite number of coefficients are non-zero, or an infinite number of coefficients are non-zero.

Definition 10.1   A point $z_0 \in \mathbb{C}$ is said to be an isolated singularity of a function f if there is $R > 0$ such that f is analytic in the punctured disc $D'(z_0, R)$ but f is not analytic at z_0.

Examples 10.1

(1) The point $z_0 = 0$ is an isolated singularity of the function $f(z) = \frac{1}{z}$. This function is not defined at 0, so is certainly not analytic there. In fact, it is not possible to assign a value at this point so that the resulting (extended) function is analytic. If this were possible, say, with $f(0)$ defined to be α, then the resulting function would have to be continuous at 0 and, in particular, bounded in some disc around 0. This is evidently false, as is seen, for example, by observing that $f(z) = n$ when $z = 1/n$.

(2) Suppose $f : \mathbb{C} \to \mathbb{C}$ is defined by $f(z) = z$ for $z \neq i$ and $f(i) = 3$. Then $z = i$ is an isolated singularity of f. (f is not even continuous at $z = i$.)

(3) Let $f(z) = 1/\sin(1/z)$. Now, $\sin(1/z) = 0$ whenever $z = 1/k\pi$ with $k \in \mathbb{Z} \setminus \{0\}$ and so f is certainly undefined at such points. However, for any given $k \in \mathbb{Z}$ with $k \neq 0$, f is analytic in the punctured disc $D'(1/k\pi, r)$ provided r is sufficiently small (depending on k). Therefore, for any $k \in \mathbb{Z}$ with $k \neq 0$, the point $z = 1/k\pi$ is an isolated singularity

of f. f is not defined for $z = 0$ and f is not analytic in any punctured disc $D'(0, r)$ for any $r > 0$ (because any such disc will contain points of the form $1/k\pi$). Evidently $z = 0$ is a singularity of f (in that f is not analytic there) but according to the definition, $z = 0$ is not an isolated singularity of f.

(4) The principal value logarithm, $f(z) = \text{Log}\, z$, is analytic in the cut-plane $\mathbb{C} \setminus \{z : z + |z| = 0\}$. Each point of the negative real axis $\{z : z + |z| = 0\}$ is a point of discontinuity of $\text{Log}\, z$ but none of these points are *isolated* singularities. Note that $\text{Log}\, z$ is not defined for $z = 0$ but is defined in the punctured plane $\mathbb{C} \setminus \{0\}$.

Let z_0 be an isolated singularity of the function f. Then, by definition, f is analytic in some punctured disc $D'(z_0, R)$. It follows that f has a Laurent expansion valid for all z in this punctured disc:

$$f(z) = \sum_{n=0}^{\infty} a_n (z - z_0)^n + \sum_{n=1}^{\infty} b_n (z - z_0)^{-n}.$$

Definition 10.2

(i) If all $b_n = 0$, then z_0 is said to be a removable singularity. (By defining or redefining f at z_0 to be equal to a_0, we get a function analytic in the whole disc $D(z_0, R)$, namely, $\sum_{n=0}^{\infty} a_n (z - z_0)^n$.)

(ii) Suppose that only a finite (but positive) number of the b_ns are non-zero; say, $b_m \neq 0$, but $b_n = 0$ for all $n > m$. Then z_0 is said to be a pole of f of order m. (Sometimes one uses the terms simple pole or double pole for the cases where $m = 1$ or $m = 2$, respectively.)

(iii) If an infinite number of the b_ns are non-zero, then z_0 is said to be an isolated essential singularity of f.

Examples 10.2

(1) Consider the function

$$f(z) = \frac{\cos z - 1}{z^2} = \frac{(1 - \frac{z^2}{2!} + \frac{z^4}{4!} - \frac{z^6}{6!} + \ldots) - 1}{z^2}$$
$$= -\frac{1}{2!} + \frac{z^2}{4!} - \frac{z^4}{6!} + \ldots$$

Evidently $z = 0$ is a removable isolated singularity.

(2) Consider

$$f(z) = \frac{\sin z}{z^3} = \frac{z - \frac{z^3}{3!} + \frac{z^5}{5!} - \cdots}{z^3}$$

$$= \frac{1}{z^2} - \frac{1}{3!} + \frac{z^2}{5!} - \cdots$$

We see that $z = 0$ is a double pole.

(3) The function

$$f(z) = \exp\frac{1}{z} = 1 + \frac{1}{z} + \frac{1}{2!\,z^2} + \frac{1}{3!\,z^3} + \cdots$$

has $z = 0$ as an essential singularity.

(4) Let

$$f(z) = \frac{(\sin z)^4}{z^4} + \cos z,$$

for $z \in \mathbb{C} \setminus \{0\}$. Then f is not defined at $z = 0$, but evidently it is bounded in, say, the punctured unit disc $D'(0,1)$. (In fact, $f(z) \to 1 + \cos 0 = 2$ as $z \to 0$.) It follows that all the b_ns vanish in the Laurent expansion of f about $z = 0$, as in proposition 9.1. Hence $z = 0$ is a removable singularity. Notice that we did not actually have to find explicitly the Laurent expansion of f about $z = 0$ to come to this conclusion.

Definition 10.3   The function f is said to be meromorphic at the point z_0 if either f is analytic at z_0 or z_0 is a pole of f. We say that the function f is meromorphic on a set $S \subseteq \mathbb{C}$ if and only if f is meromorphic at each point in S.

10.2   Behaviour near an Isolated Singularity

Theorem 10.1   *Suppose that $f \in H(D'(z_0, R))$. Then z_0 is a pole of order m if and only if the limit $\lim_{z \to z_0}(z - z_0)^m f(z)$ exists and is non-zero.*

Proof.   Suppose first that z_0 is pole of f of order m. This means that b_m is the last non-zero coefficient (of the power $(z - z_0)^{-m}$) in the principal part of the Laurent expansion of f about z_0 ($b_n = 0$ for all $n > m$). Evidently,

$$\lim_{z \to z_0} (z - z_0)^m f(z) = b_m \neq 0,$$

as required.

Conversely, suppose that $\lim_{z \to z_0} (z - z_0)^m f(z) = \alpha \neq 0$. Let F be given by $F(z) = (z - z_0)^m f(z)$ for $z \in D'(z_0, R)$. Since f and therefore F is analytic in this punctured disc, it follows that F has a Laurent expansion

$$F(z) = \sum_{n=0}^{\infty} A_n (z - z_0)^n + \sum_{n=1}^{\infty} B_n (z - z_0)^{-n}$$

valid (absolutely convergent) in $D'(z_0, R)$. However, by hypothesis, we have that $F(z) \to \alpha$ as $z \to z_0$ and so F is bounded in the neighbourhood of z_0, say, in $D'(z_0, r)$. ($|F(z)|$ is close to $|\alpha|$ if z is sufficiently close to z_0.) It follows that $B_n = 0$ for all $n \in \mathbb{N}$ (as in proposition 9.1).
Hence $F(z) = \sum_{n=0}^{\infty} A_n (z - z_0)^n$ and

$$0 \neq \alpha = \lim_{z \to z_0} F(z) = A_0.$$

Therefore

$$(z - z_0)^m f(z) = A_0 + A_1(z - z_0) + A_2(z - z_0)^2 + \dots$$

so that

$$f(z) = \frac{A_0}{(z - z_0)^m} + \frac{A_1}{(z - z_0)^{m-1}} + \dots + A_m + A_{m+1}(z - z_0) + \dots$$

for any $z \in D'(z_0, R)$. Since $A_0 = \alpha \neq 0$, we see that z_0 is a pole of f of order m. $\qquad \square$

Example 10.3   The function $f(z) = \dfrac{1}{e^z - 1}$ is undefined at those points z for which $e^z = 1$, i.e., when $z = 2k\pi i$ for $k \in \mathbb{Z}$. f is analytic everywhere else and so these are isolated singularities. Fix $k \in \mathbb{Z}$ and set $w = z - 2k\pi i$. Then $f(z) = f(w + 2k\pi i) = 1/(e^{w+2k\pi i} - 1) = 1/(e^w - 1)$. Using the power series expansion of e^w, we see that $(e^w - 1)/w \to 1$ as $w \to 0$ and so we deduce that $(z - 2k\pi i) f(z) \to 1 \neq 0$ as $z \to 2k\pi i$. It follows that for each $k \in \mathbb{Z}$, $z = 2k\pi i$ is a simple pole of f.

Proposition 10.1   *Suppose z_0 is an isolated singularity of f and suppose that $|f(z)| \to \infty$ as $z \to z_0$. Then z_0 is a pole of f (so f is meromorphic at z_0). Conversely, if z_0 is a pole of f, then $|f(z)| \to \infty$ as $z \to z_0$.*

Proof.   Suppose $|f(z)| \to \infty$ as $z \to z_0$. Then, for any given $M > 0$, there is $R > 0$ such that $f \in H(D'(z_0, R))$ and $|f(z)| > M$ for all $z \in D'(z_0, R)$. In particular, $f(z) \neq 0$ on the punctured disc $D'(z_0, R)$. It follows that $g(z) = 1/f(z) \in H(D'(z_0, R))$ and g satisfies $|g(z)| < 1/M$ on $D'(z_0, R)$.

By proposition 9.1, we see that z_0 is a removable singularity of g and that g can be written as (using $\lim_{z \to z_0} g(z) = 0$)

$$g(z) = A_1(z - z_0) + A_2(z - z_0)^2 + \ldots$$

for any z in $D'(z_0, R)$. Furthermore, g is not zero in this punctured disc (because $g(z)f(z) = 1$) and therefore not all of the A_ns are zero. That is, there is $m \geq 1$ such that $A_m \neq 0$ but $A_n = 0$ for all $n < m$. It follows that

$$\frac{1}{f(z)} = g(z) = (z - z_0)^m \left(A_m + A_{m+1}(z - z_0) + \ldots \right)$$

and so

$$\frac{1}{(z - z_0)^m f(z)} \to A_m$$

as $z \to z_0$, i.e., $(z - z_0)^m f(z) \to \dfrac{1}{A_m} \neq 0$. By the theorem, theorem 10.1, it follows that z_0 is a pole of f of order m.

For the converse, suppose that z_0 is a pole of f of order $m \geq 1$, say. Then, by theorem 10.1, $(z - z_0)^m f(z) \to \alpha \neq 0$ as $z \to z_0$. Hence, for $z \neq z_0$ (and in some neighbourhood of z_0 so that $f(z)$ is defined)

$$\left| \frac{1}{f(z)} \right| = \left| \frac{(z - z_0)^m}{(z - z_0)^m f(z)} \right| \to \left| \frac{0}{\alpha} \right| = 0$$

as $z \to z_0$ and the result follows. □

Corollary 10.1 *Suppose that z_0 is an isolated essential singularity of f. Then f is neither bounded near z_0 nor does $|f(z)|$ diverge to ∞ as $z \to z_0$. In other words, there are sequences (z_n) and (ζ_n) such that both $z_n \to z_0$ and $\zeta_n \to z_0$ and such that $|f(z_n)| \to \infty$ as $n \to \infty$ but $(f(\zeta_n))$ is bounded.*

Proof. If f were bounded near z_0, then z_0 would be a removable isolated singularity. On the other hand, if $|f(z)| \to \infty$ as $z \to z_0$, then we have seen that this would imply that z_0 is a pole of f. By hypothesis, neither of these possibilities hold and so the result follows. □

What's going on?   To say that $\lim_{z \to z_0} (z - z_0)^m f(z) = \alpha \neq 0$ means that one can think of $f(z)$ as behaving rather like $\alpha/(z - z_0)^m$ for all z near to z_0. It seems quite reasonable that this amounts z_0 being a pole of f of order m. Also, if z_0 is a pole of f, then we would expect that $|f(z)| \to \infty$ as $z \to z_0$. However, the converse, although true (as we have seen), is not quite so obvious. After all, one could be forgiven for imagining that it was possible for f to have an infinite number of terms in the principal part of its Laurent expansion in some punctured disc around z_0 (and so z_0 would be an essential singularity) and *still* be such that $|f(z)| \to \infty$ as $z \to z_0$. That this cannot happen is perhaps something of a surprise.

Example 10.4   Define $f(z)$, for any $z \neq 0$, by $f(z) = \cos \frac{1}{z}$. Then, for any $z \neq 0$,

$$f(z) = \cos \frac{1}{z} = 1 - \frac{1}{2! \, z^2} + \frac{1}{4! \, z^4} - \cdots$$

and we see that $z_0 = 0$ is an essential singularity of f. Taking $z_n = i/n$ and $\zeta_n = 1/n$, for $n \in \mathbb{N}$, we see that $z_n \to 0$ and $\zeta_n \to 0$ as $n \to \infty$ but $f(z_n) = \cos(n/i) = \frac{1}{2}(e^n + e^{-n}) \to \infty$ as $n \to \infty$, whereas the sequence $(f(\zeta_n)) = (\cos n)$ is bounded.

10.3   Behaviour as $|z| \to \infty$

Theorem 10.2   *Let f be entire and non-constant. Then f is a polynomial if and only if $|f(z)| \to \infty$ as $|z| \to \infty$.*

Proof.   Suppose, first, that f is a (non-constant) polynomial,

$$f(z) = a_n z^n + \cdots + a_1 z + a_0$$

with $a_n \neq 0$. Then

$$\left| \frac{1}{f(z)} \right| = \left| \frac{1}{a_n z^n + \cdots + a_0} \right|$$
$$= \left| \frac{1}{z^n \left(a_n + \frac{a_{n-1}}{z} + \cdots + \frac{a_0}{z^n} \right)} \right| \to 0,$$

as $|z| \to \infty$.

Conversely, suppose that $|f(\zeta)| \to \infty$ as $|\zeta| \to \infty$. Since f is entire, it has a Taylor expansion (about $z = 0$)

$$f(z) = \sum_{n=0}^{\infty} a_n z^n, \quad \text{for all } z.$$

For $z \neq 0$, set

$$g(z) = f\left(\frac{1}{z}\right) = \sum_{n=0}^{\infty} a_n z^{-n}.$$

Then $g \in H(\mathbb{C} \setminus \{0\})$. Moreover, this series converges absolutely for $|z| > 0$ and so is *the* Laurent expansion of g (about $z = 0$). Now, as $z \to 0$, $\left|\frac{1}{z}\right| \to \infty$ and so, by hypothesis, $\left|f(\frac{1}{z})\right| \to \infty$, i.e., $|g(z)| \to \infty$ as $z \to 0$.

It follows that $z = 0$ is a pole of g and therefore

$$g(z) = a_0 + \frac{a_1}{z} + \cdots + \frac{a_m}{z^m}$$

for some $m \geq 1$, where $a_m \neq 0$. Hence

$$f(z) = g\left(\frac{1}{z}\right) = a_0 + a_1 z + \cdots + a_m z^m$$

for $z \neq 0$. However, from the Taylor expansion of f above, we see that $f(0) = a_0$ and so

$$f(z) = a_0 + a_1 z + \cdots + a_m z^m$$

for all $z \in \mathbb{C}$ (including $z = 0$). That is, f is a polynomial. $\quad\square$

Corollary 10.2 *Any entire non-polynomial function f has the property that there is a sequence (z_n) such that $|z_n| \to \infty$ and $|f(z_n)| \to \infty$ as $n \to \infty$ and another sequence (ζ_n), say, such that $|\zeta_n| \to \infty$ as $n \to \infty$ but the sequence $(f(\zeta_n))$ is bounded.*

Proof. First we note that by Liouville's Theorem, theorem 8.10, since f is entire it cannot be bounded, unless it is a constant. However, it cannot be constant because it is not a polynomial. The existence of a sequence (z_n), as above, then follows.

Furthermore, again because f is not a polynomial, the existence of some sequence (ζ_n), as above, follows from the previous theorem. $\quad\square$

What's going on ? It is intuitively clear that if f is a polynomial then $|f(z)|$ must diverge to ∞ as $|z| \to \infty$. It is far from obvious that the converse is true. After all, one might suspect that a non-polynomial function such as $e^z + e^{iz}$, or something similar, would exhibit this behaviour. Not so. (In this particular example, set $z = it - t$ for $t \in \mathbb{R}$ and $t > 0$. Then $e^z + e^{iz} = e^{it}e^{-t} + e^{-t}e^{-it} = 2e^{-t}\cos t$ which stays bounded as $t \to \infty$.)

10.4   Casorati-Weierstrass Theorem

The next result tells us that near an isolated essential singularity a function takes values arbitrarily close to any given complex number.

Theorem 10.3 (Casorati-Weierstrass Theorem)   *Suppose z_0 is an isolated essential singularity of the function f. Let $r > 0$, $\varepsilon > 0$ and $w \in \mathbb{C}$ be given. Then there is some z with $|z - z_0| < r$ such that $|f(z) - w| < \varepsilon$.*

Proof.   We know that there is some $\rho > 0$ such that f is analytic in the punctured disc $D'(z_0, \rho)$. Let $r' = \min\{r, \rho\}$.

By way of contradiction, suppose that $|f(z) - w| \geq \varepsilon$ for all $z \in D'(z_0, r')$. Then it follows that $g(z) = 1/(f(z) - w)$ is analytic and bounded in $D'(z_0, r')$. This means that z_0 is a removable singularity of g, i.e., we can extend g to an analytic function in the whole disc $D(z_0, r')$. Let us denote this extension by G. Then

$$G(z)(f(z) - w) = 1$$

for $z \in D'(z_0, r')$. If $G(z_0) \neq 0$, this entails that f be bounded near z_0.

On the other hand, if $G(z_0) = 0$ then $|f(z)|$ must diverge as $z \to z_0$. The first case would mean that z_0 were a removable singularity and the second that it was a pole of f. However, z_0 is neither of these, it is an essential singularity of f. We conclude that $|f(z) - w| \geq \varepsilon$ for all $z \in D'(z_0, r')$ is false, and the result follows.   □

Remark 10.1   In fact, it has been shown that a function takes on all values with at most one exception in any neighbourhood of an essential singularity, but this is harder to prove (Picard's Theorem). The example $\exp(1/z)$ shows that there may be an exception. Evidently $z_0 = 0$ is an isolated essential singularity but the exponential function is never zero.

Chapter 11

Theory of Residues

11.1 Residues

The function z^n has a primitive on $\mathbb{C}$ if $n \geq 0$ and has a primitive on $\mathbb{C} \setminus \{0\}$ if $n < -1$. This means that the integral of the function z^n around any closed contour (not passing through 0 if $n < -1$) always vanishes, except for the single case, $n = -1$. For this reason, the $n = -1$ term in the Laurent expansion of a function plays a special rôle.

Definition 11.1 Suppose that f is meromorphic in a set S and let $z_0 \in S$ so that, for some $R > 0$,

$$f(z) = \sum_{n=0}^{\infty} a_n(z - z_0)^n + \sum_{n=1}^{m} b_n(z - z_0)^{-n}$$

for $z \in D'(z_0, R)$, and where m possibly depends on z_0. The residue of f at z_0 is defined as $\mathrm{Res}(f : z_0) = b_1$. Thus,

$$\mathrm{Res}(f : z_0) = b_1 = \frac{1}{2\pi i} \int_\gamma f \,,$$

where $\gamma(t) = z_0 + re^{2\pi it}$, $0 \leq t \leq 1$, for any $0 < r < R$.

Theorem 11.1 *Suppose that z_0 is a pole of f of order m. Then*

$$\mathrm{Res}(f : z_0) = \frac{1}{(n-1)!} \lim_{z \to z_0} \frac{d^{n-1}}{dz^{n-1}} \Big((z - z_0)^n f(z) \Big)$$

for any $n \geq m$.

Proof. We have

$$f(z) = \sum_{k=0}^{\infty} a_k(z - z_0)^k + \frac{b_1}{(z - z_0)} + \cdots + \frac{b_m}{(z - z_0)^m}$$

and so

$$(z - z_0)^n f(z) = (z - z_0)^n \sum_{k=0}^{\infty} a_k(z - z_0)^k + b_1(z - z_0)^{n-1} + \ldots$$

$$\ldots + b_m(z - z_0)^{n-m}$$

for any $n \geq m$. The result follows directly. $\square$

Remark 11.1 Usually, one takes $n = m$ to find the residue. However, it is sometimes more convenient to choose a suitable $n > m$. Consider, for example, the function $f(z) = \dfrac{\sin z}{z^4}$. The point $z = 0$ is a pole of order 3. Taking $n = m = 3$, we find

$$\mathrm{Res}(f:0) = \frac{1}{2!} \lim_{z \to 0} \frac{d^2}{dz^2}\left(\frac{z^3 \sin z}{z^4}\right)$$

which requires calculation of $\dfrac{d^2}{dz^2}\left(\dfrac{\sin z}{z}\right)$.

On the other hand, taking $n = 4 > 3 = m$, we get

$$\mathrm{Res}(f:0) = \frac{1}{3!} \lim_{z \to 0} \frac{d^3}{dz^3}\left(\frac{z^4 \sin z}{z^4}\right)$$
$$= \frac{1}{3!}(-\cos z)\big|_{z=0} = -\frac{1}{6}$$

which is marginally easier to calculate.

Example 11.1 The function $\sec z = 1/\cos z$ has isolated singularities when $\cos z = 0$, namely, when z is of the form $z = (2n - 1)\frac{\pi}{2}$ for $n \in \mathbb{Z}$. Put $w = z - k\frac{\pi}{2}$ where $k = 2n - 1$. Then

$$\sec z = \sec(w + k\tfrac{\pi}{2}) = \frac{1}{\{\cos w \cos k\frac{\pi}{2} - \sin w \sin k\frac{\pi}{2}\}}$$
$$= \frac{-1}{\sin w \sin k\frac{\pi}{2}} = \frac{(-1)^n}{\sin w}.$$

Since $w/\sin w \to 1$ as $w \to 0$, we deduce that $(z - (2n-1)\frac{\pi}{2})\sec z \to (-1)^n$ as $z \to (2n - 1)\frac{\pi}{2}$. It follows that these are all simple poles, with residues $(-1)^n$, respectively.

11.2 Winding Number (Index)

Let $\gamma(t) = z_0 + r\,e^{2\pi i k t}$ for $t \in [0,1]$ and $k \in \mathbb{N}$. Then we find that

$$\frac{1}{2\pi i} \int_\gamma \frac{dz}{z - z_0} = k\,,$$

the number of times the circular path γ "goes around" the point z_0. By deformation, we might expect that this holds even if γ is changed slightly so that it is no longer circular but still goes round z_0 k times. In fact, we can use the integral formula to tell us how many times γ does encircle the point z_0. This leads to the notion of winding number for any contour (not necessarily circular), as we discuss next.

Theorem 11.2 *For any closed contour γ and any point $z_0 \notin \operatorname{tr} \gamma$,*

$$\frac{1}{2\pi i} \int_\gamma \frac{dz}{z - z_0} \in \mathbb{Z}.$$

Proof. Suppose first that $\gamma : [a,b] \to \mathbb{C}$ is a smooth path. Let

$$h(t) = \int_a^t \frac{\gamma'(s)}{(\gamma(s) - z_0)}\,ds$$

for $a \le t \le b$. Then $h'(t) = \dfrac{\gamma'(t)}{(\gamma(t) - z_0)}$ on (a,b), since the integrand is continuous. Hence

$$\frac{d}{dt}\Big((\gamma(t) - z_0)\exp(-h(t))\Big) = \gamma'(t)\,e^{-h(t)} - (\gamma(t) - z_0)\,h'(t)\,e^{-h(t)}$$

$$= 0.$$

We deduce that $(\gamma(t) - z_0)\,e^{-h(t)}$ is constant on $[a,b]$.

For a general contour $\gamma : [a,b] \to \mathbb{C}$, we may write $\gamma = \gamma_1 + \gamma_2 + \cdots + \gamma_m$ for $a = t_0 < t_1 < \cdots < t_m = b$ and smooth sub-paths $\gamma_j : [t_{j-1}, t_j] \to \mathbb{C}$, $1 \le j \le m$. The function h is now defined by

$$h(s) = \begin{cases} \int_a^s \frac{\gamma'(s)}{(\gamma(s) - z_0)}\,ds\,, & \text{for } a \le s \le t_1, \\[2mm] h(t_{j-1}) + \int_{t_{j-1}}^s \frac{\gamma'(s)}{(\gamma(s) - z_0)}\,ds\,, & \text{for } t_{j-1} < s \le t_j \text{ and } 2 \le j \le m. \end{cases}$$

Then h is continuous on $[a, b]$ and, as above, $(\gamma(t) - z_0)\, e^{-h(t)}$ is constant on each interval (t_{j-1}, t_j) and therefore on the whole of $[t_0, t_m] = [a, b]$. Hence, equating values at the end-points, we get

$$(\gamma(b) - z_0)\, e^{-h(b)} = (\gamma(a) - z_0)\, e^{-h(a)}.$$

It follows that $e^{-h(b)} = e^{-h(a)}$, since $\gamma(a) = \gamma(b)$ (because γ is closed) and $z_0 \notin \operatorname{tr}\gamma$. Hence $h(b) = h(a) + 2\pi k i$ for some $k \in \mathbb{Z}$. But $h(a) = 0$, and $h(b) = \displaystyle\int_\gamma \frac{dz}{z - z_0}$ and so

$$\int_\gamma \frac{dz}{z - z_0} = 2\pi k i$$

for some $k \in \mathbb{Z}$, as required.   □

If $\operatorname{tr}\gamma$ is a circle centred on z_0, then we know that k is just the number of times γ goes around the point z_0. This suggests the terminology of "winding number" or "index" of a (closed) contour with respect to a given point not on the contour.

Definition 11.2   For any closed contour γ and any point $z_0 \notin \operatorname{tr}\gamma$, the integer $\dfrac{1}{2\pi i}\displaystyle\int_\gamma \dfrac{dz}{z - z_0}$ is called the winding number (or index) of γ with respect to z_0, and is denoted by $\operatorname{Ind}(\gamma : z_0)$.

Examples 11.2

(1) For $\gamma(t) = re^{2\pi i t}$, $0 \le t \le 1$, we find that $\operatorname{Ind}(\gamma:0) = 1$.
(2) If $\gamma(t) = re^{-6\pi i t}$, $0 \le t \le 1$, then $\operatorname{Ind}(\gamma:0) = -3$.
(3) For $\gamma(t) = e^{4\pi i t}$, $0 \le t \le 2$, $\operatorname{Ind}(\gamma:0) = 4$, $\operatorname{Ind}(\gamma:3i) = 0$.
(4) Deformation techniques can be used to determine winding numbers. For example, let γ be the simple closed contour whose trace is the diamond with vertices $\pm i$ and ± 1. The function $1/z$ is analytic in the star-domain $D = \{z : z = re^{i\theta},\ r > 0,\ -\frac{1}{4}\pi < \theta < \frac{3}{4}\pi\}$ and so it follows that

$$\int_{[1,i]} \frac{dz}{z} = \int_\varphi \frac{dz}{z}$$

where φ is the quarter-circle $\varphi(t) = e^{it}$, $0 \le t \le \frac{1}{2}\pi$. Applying this idea to the other sides of the square (in the suitably rotated star-domains,

$e^{i\pi/2}D$, $e^{i\pi}D$ and $e^{i3\pi/2}$) and adding the results, we see that

$$\int_\gamma \frac{dz}{z} = \int_\psi \frac{dz}{z}$$

where ψ is the circle $\psi(t) = e^{it}$, $0 \le t \le 2\pi$. Hence $\mathrm{Ind}(\gamma{:}0) = 1$, as it should.

11.3 Cauchy's Residue Theorem

The next theorem is yet another fundamental result.

Theorem 11.3 (Cauchy's Residue Theorem) *Let f be a meromorphic function with a finite number of poles, $\zeta_1, \ldots, \zeta_m$, say, in the star-domain D and let γ be a closed contour with $\mathrm{tr}\,\gamma \subseteq D\backslash\{\zeta_1, \ldots, \zeta_m\}$ (i.e., the contour does not pass through any of the poles). Then*

$$\int_\gamma f = 2\pi i \sum_{k=1}^m \mathrm{Res}(f{:}\zeta_k)\,\mathrm{Ind}(\gamma{:}\zeta_k).$$

Proof. Suppose that ζ_k is a pole of order m_k. Then there is $R_k > 0$ such that

$$f(z) = \sum_{n=0}^\infty a_n^{(k)}(z - \zeta_k)^n + P_k(z)$$

for $z \in D'(\zeta_k, R_k)$, where $P_k(z) = \sum_{n=1}^{m_k} b_n^{(k)}(z - \zeta_k)^{-n}$. The function P_k is analytic in $\mathbb{C} \setminus \{\zeta_k\}$ and by defining $f - P_k$ to be equal to $a_0^{(k)}$ at ζ_k, we see that $f - P_k$ is analytic at ζ_k. Put

$$g(z) = f(z) - \sum_{k=1}^m P_k(z).$$

Then $g \in H(D)$. Note that $g(\zeta_j)$ is defined to be

$$g(\zeta_j) = (f - P_j)(\zeta_j) - \sum_{k \ne j}^m P_k(\zeta_j) = a_0^{(j)} - \sum_{k \ne j}^m P_k(\zeta_j).$$

Since D is star-like, it follows that $\int_\gamma g = 0$. Hence

$$\int_\gamma f = \sum_{k=1}^m \int_\gamma P_k.$$

But

$$\int_\gamma P_k = \int_\gamma \sum_{n=1}^{m_k} b_n^{(k)} (z - \zeta_k)^{-n} \, dz = b_1^{(k)} \int_\gamma \frac{dz}{z - \zeta_k}$$

$$= 2\pi i \, b_1^{(k)} \, \mathrm{Ind}(\gamma : \zeta_k)$$

and the result follows.   □

Example 11.3   We will show that

$$\int_\gamma \frac{\cos z}{z(z-1)^2} \, dz = 2\pi i (1 - \sin 1 - \cos 1),$$

where γ is the circle $\gamma(t) = 3e^{2\pi i t}$, $0 \le t \le 1$. To see this, first note that the singularities of the integrand $f(z) = \cos z / z(z-1)^2$ are at $z = 0$ and $z = 1$. Clearly, $\mathrm{Ind}(\gamma : 0) = 1 = \mathrm{Ind}(\gamma : 1)$. Therefore

$$\frac{1}{2\pi i} \int_\gamma f = \mathrm{Res}(f : 0) + \mathrm{Res}(f : 1).$$

Now,

$$\mathrm{Res}(f : 0) = \lim_{z \to 0} \frac{z \cos z}{z(z-1)^2} = 1$$

and

$$\mathrm{Res}(f : 1) = \lim_{z \to 1} \frac{d}{dz}\left((z-1)^2 \frac{\cos z}{z(z-1)^2} \right)$$

$$= \lim_{z \to 1} \left(-\frac{\sin z}{z} - \frac{\cos z}{z^2} \right)$$

$$= -\sin 1 - \cos 1.$$

Example 11.4   Suppose that f is analytic in $\mathbb{C}$ except for poles at the points ± 1 where it has residues $\mathrm{Res}(f : -1) = 2$ and $\mathrm{Res}(f : 1) = 5$, respectively. Let $\gamma : [0,1] \to \mathbb{C}$ be the contour

$$\gamma(t) = \begin{cases} -1 + e^{12 i \pi t}, & \text{for } 0 \le t < \tfrac{1}{2} \\ 1 + e^{i\pi(1-8t)}, & \text{for } \tfrac{1}{2} \le t \le 1. \end{cases}$$

Then γ winds 3 times around $z = -1$ and winds around $z = 1$ twice, but in the negative sense. Therefore

$$\frac{1}{2\pi i} \int_\gamma f = 2 \times 3 + 5 \times (-2) = -4.$$

The next example indicates a way of performing real integrals involving trigonometric functions—courtesy of the relationship $e^{it} = \cos t + i \sin t$.

Example 11.5

$$\int_0^{2\pi} \frac{dt}{\sqrt{5} + \cos t} = \pi.$$

To show how such integrals might be evaluated, notice first that an integral from 0 to 2π suggests an integral around a circle. Let us write the integrand in terms of e^{it}. We have $\cos t = \frac{1}{2}(e^{it} + e^{-it})$. Let $\gamma(t) = e^{it}$, $0 \le t \le 2\pi$. Then $\cos t = \frac{1}{2}(\gamma(t) + 1/\gamma(t))$ and $\gamma'(t) = ie^{it} = i\gamma(t)$ so the integral becomes

$$\int_0^{2\pi} \frac{1}{\sqrt{5} + \frac{1}{2}(\gamma(t) + \gamma(t)^{-1})} \frac{\gamma'(t)}{i\gamma(t)} dt = \int_\gamma \frac{1}{\sqrt{5} + \frac{1}{2}(z + z^{-1})} \frac{1}{iz} dz$$

$$= 2\pi i \sum_{|\zeta_k|<1} \text{Res}(f:\zeta_k)$$

where $f = \dfrac{1}{\{\sqrt{5} + \frac{1}{2}(z + z^{-1})\} iz}$ and the sum is over those poles ζ_k of f which lie inside the unit circle (because $\text{Ind}(\gamma:\zeta) = 0$ for any pole ζ of f outside this circle). Notice that there are no poles *on* the circle tr γ because $\sqrt{5} + \cos t$ is never zero (for t real). To find the poles of f, we write

$$f(z) = \frac{1}{i(z\sqrt{5} + \frac{1}{2}(z^2 + 1))}$$

$$= \frac{2}{i(z^2 + 2\sqrt{5}\,z + 1)}$$

$$= \frac{2}{i(z + \sqrt{5} - 2)(z + \sqrt{5} + 2)}$$

and so we see that f has poles at $z = -\sqrt{5} \pm 2$. Now, $\text{Ind}(\gamma:-\sqrt{5} - 2) = 0$ because $-\sqrt{5} - 2$ is outside γ and $\text{Ind}(\gamma:-\sqrt{5} + 2) = 1$. Therefore,

$$\int_0^{2\pi} \frac{dt}{\sqrt{5} + \cos t} = 2\pi i\,\text{Res}(f:-\sqrt{5} + 2)$$

$$= 2\pi i\,\frac{2}{i}\lim_{z \to -\sqrt{5}+2} \frac{(z + \sqrt{5} + 2)}{(z + \sqrt{5} - 2)(z + \sqrt{5} + 2)}$$

$$= 2\pi i\,\frac{2}{i}\frac{1}{4}$$

$$= \pi.$$

Theorem 11.4   *Suppose that f is analytic in $\{\, z : \operatorname{Im} z \geq 0 \,\}\backslash\{z_1, \ldots, z_n\}$ with poles at the points $z_1, \ldots, z_n$ in the upper half-plane (i.e., $\operatorname{Im} z_k > 0$, $1 \leq k \leq n$). Suppose further that $zf(z) \to 0$ whenever $|z| \to \infty$ with $\operatorname{Im} z \geq 0$. Then*

$$\lim_{R \to \infty} \int_{-R}^{R} f(x)\,dx = 2\pi i \sum_{k=1}^{n} \operatorname{Res}(f : z_k).$$

Proof.   Let $\varepsilon > 0$ be given. By hypothesis, there is R so that $|zf(z)| < \varepsilon$ for all $|z| > R$ with $\operatorname{Im} z \geq 0$. Let Σ be the semicircular path with centre 0 and radius ρ and let $\Gamma = [-\rho, \rho] + \Sigma$. Choose ρ so large that Γ encloses all the poles $z_1, \ldots, z_n$. Then $\operatorname{Ind}(\Gamma : z_k) = 1$ for all $1 \leq k \leq n$ and so, by the Residue Theorem,

$$\int_{\Gamma} f = 2\pi i \sum_{k=1}^{n} \operatorname{Res}(f : z_k).$$

Now,

$$\int_{\Gamma} f = \int_{[-\rho, \rho]} f + \int_{\Sigma} f$$

and we claim that $\int_{\Sigma} f \to 0$ as $\rho \to \infty$. To see this, suppose $\rho > R$. Then, for any $z \in \operatorname{tr}\Sigma$, $|zf(z)| < \varepsilon$, that is, $\rho\,|f(z)| < \varepsilon$, since $|z| = \rho$ for $z \in \operatorname{tr}\Sigma$. Hence $|f(z)| < \varepsilon/\rho$ for all $z \in \operatorname{tr}\Sigma$. By the Basic Estimate, we get

$$\left| \int_{\Sigma} f \right| \leq \frac{\varepsilon}{\rho} L(\Sigma) = \frac{\varepsilon}{\rho} \pi\rho = \varepsilon\pi$$

and it follows that $\int_{\Sigma} f \to 0$ as $\rho \to \infty$, as claimed.

The result now follows because $\int_{[-\rho,\rho]} f = \int_{-\rho}^{\rho} f(x)\,dx.$ □

Note that in the theorem above, there are no poles *on* the real axis.

Example 11.6   We evaluate $\displaystyle\int_{-\infty}^{\infty} \frac{dx}{(x^2 + 1)(x^2 + 4)(x^2 + 9)}.$

To do this, let $f(z) = \dfrac{1}{(z^2 + 1)(z^2 + 4)(z^2 + 9)}$. Evidently, f satisfies the hypotheses of the theorem. The poles of f are at $\pm i$, $\pm 2i$ and $\pm 3i$ and are all simple poles. By the theorem,

$$\int_{-\infty}^{\infty} \frac{dx}{(x^2+1)(x^2+4)(x^2+9)} = \lim_{\rho\to\infty} \int_{-\rho}^{\rho} \frac{dx}{(x^2+1)(x^2+4)(x^2+9)}$$

$$= 2\pi i \sum_{\text{poles } \zeta \text{ in } \{\,\text{Im } z > 0\,\}} \text{Res}(f:\zeta)$$

$$= 2\pi i \left(\text{Res}(f:i) + \text{Res}(f:2i) + \text{Res}(f:3i) \right).$$

All we now need do is to calculate the residues. For example,

$$\text{Res}(f:i) = \lim_{z\to i}(z-i)f(z) = \frac{1}{2i \times 3 \times 8}\,.$$

The other two residues can be similarly calculated.

Chapter 12

The Argument Principle

12.1 Zeros and Poles

We shall see that the number of zeros and poles of a meromorphic function f is determined by the behaviour of the quotient $\frac{f'}{f}$.

Theorem 12.1 *Suppose that f is analytic at z_0 and that z_0 is a zero of f of order m. Then $\frac{f'}{f}$ is meromorphic at z_0 and $\operatorname{Res}(\frac{f'}{f} : z_0) = m$. In fact, the point z_0 is a simple pole of $\frac{f'}{f}$.*

Proof. By hypothesis, we can write

$$f(z) = \sum_{n=m}^{\infty} a_n (z - z_0)^n$$

where $a_m \neq 0$ and this power series converges absolutely for all z in some disc $D(z_0, R)$. Since $a_m \neq 0$, it follows that f is not identically zero in $D(z_0, R)$ and so there is $r > 0$ such that $f(z) \neq 0$ for all z in the punctured disc $D'(z_0, r)$ (otherwise f would vanish, by the Identity Theorem). It follows that $1/f$ is analytic in $D'(z_0, r)$ and so, therefore, is f'/f. Hence, z_0 is an isolated singularity of f'/f. Write

$$f(z) = (z - z_0)^m \underbrace{\left(a_m + a_{m+1}(z - z_0) + \ldots\right)}_{\varphi(z)}.$$

for $z \in D(z_0, R)$. Differentiating, we get

$$f'(z) = m(z - z_0)^{m-1}\varphi(z) + (z - z_0)^m \varphi'(z).$$

185

Now, $\varphi(z_0) = a_m \neq 0$ and so, by continuity, there is $\rho > 0$ such that $\varphi(z) \neq 0$ for all $z \in D(z_0, \rho)$. Hence, for any $z \in D'(z_0, \rho)$, we may write

$$\frac{f'(z)}{f(z)} = \frac{m}{(z - z_0)} + \frac{\varphi'(z)}{\varphi(z)}.$$

Furthermore, φ'/φ is analytic in the disc $D(z_0, \rho)$ and so has a Taylor series expansion there. Therefore

$$\frac{f'(z)}{f(z)} = \frac{m}{(z - z_0)} + \sum_{n=0}^{\infty} A_n(z - z_0)^n$$

for suitable coefficients (A_n), where the series converges absolutely in $D(z_0, \rho)$. This is the Laurent expansion of f'/f at z_0 and the result follows. $\square$

There is a corresponding result for poles.

Theorem 12.2 *Suppose that z_0 is a pole of f of order k. Then $\frac{f'}{f}$ has a simple pole at z_0 with $\mathrm{Res}(\frac{f'}{f} : z_0) = -k$.*

Proof. By hypothesis, f has the Laurent expansion

$$f(z) = \frac{b_k}{(z - z_0)^k} + \cdots + \sum_{n=0}^{\infty} a_n(z - z_0)^n,$$

where $b_k \neq 0$, valid in some punctured disc $D'(z_0, R)$. Hence we may write

$$f(z) = \frac{\psi(z)}{(z - z_0)^k}$$

where ψ is analytic at z_0 (in fact, in $D(z_0, R)$) and $\psi(z_0) = b_k \neq 0$. By continuity, it follows that ψ is non-zero in some neighbourhood of z_0 and so there is some $\rho > 0$ such that ψ is analytic and does not vanish in the disc $D(z_0, \rho)$. From the equality above, we see that f does not vanish in the punctured disc $D'(z_0, \rho)$. Hence

$$\frac{f'(z)}{f(z)} = \frac{-k}{(z - z_0)} + \frac{\psi'(z)}{\psi(z)}$$

for $z \in D'(z_0, \rho)$. The last term ψ'/ψ is analytic in this disc (so has a Taylor expansion) and we deduce that z_0 is a simple pole of f'/f and that $\mathrm{Res}(f'/f : z_0) = -k$, as claimed. $\square$

12.2　Argument Principle

Applying these results, we obtain the following theorem.

Theorem 12.3 (Argument Principle) *Suppose that D is a star-domain and γ is a contour in D such that $\mathrm{Ind}(\gamma : \zeta) = 0$ or 1 for any $\zeta \notin \mathrm{tr}\,\gamma$. Suppose, further, that f is meromorphic in D, has a finite number of poles in D and is such that none of its poles nor zeros lie on $\mathrm{tr}\,\gamma$. Then*

$$\frac{1}{2\pi i}\int_\gamma \frac{f'}{f} = N_\gamma - P_\gamma$$

where N_γ is the number of zeros of f inside γ (counting multiplicity) and P_γ is the number of poles of f inside γ (counting multiplicity).

Proof. This is an immediate consequence of the previous two theorems together with the Residue Theorem, theorem 11.3. □

Remark 12.1 It is called the Argument Principle for the following reason. The expression $\frac{1}{2\pi i}\int_\gamma f'/f$ looks as though it should be just $\frac{1}{2\pi i}\log f\big]_\gamma$ but we have seen that the logarithm must be handled with care. To make some sense of this, suppose that we break the contour γ into a number of (possibly very small) subcontours, γ_k. The integral is then the sum of the integrals along these subcontours. Since f does not vanish on $\mathrm{tr}\,\gamma$, we can imagine that each subcontour is contained in some disc also on which f does not vanish. (This is a consequence of the continuity of f.) We can also imagine that these discs are so small (if necessary) that the values $f(z)$ taken by f on each such disc lie inside some disc in $\mathbb{C}$ which does not contain 0. This means that there is a branch of the logarithm defined on the values $f(z)$, as z varies on a given subcontour, γ_k. Now the integral really is the logarithm

$$\frac{1}{2\pi i}\int_{\gamma_k} \frac{f'}{f} = \log f(z_k) - \log f(z_{k-1})$$

$$= \log|f(z_k)| + i \arg f(z_k) - \log|f(z_{k-1})| - i \arg f(z_{k-1})$$

$$= \log|f(z_k)| - \log|f(z_{k-1})| + i\Delta_k \arg f,$$

where $\log$ denotes the branch constructed above, z_{k-1} and z_k are the ends of γ_k and $\Delta_k \arg f$ denotes the variation of the argument of $f(z)$ as z moves along γ_k.

The idea is to do this for each of the subcontours, but being careful to ensure that the choices of branch of logarithm (or argument) match up at the values of f at the ends of the γ_ks.

Integrating along γ means adding up these subintegrals and we see that the $\log |f(z)|$ terms cancel out leaving just the sum

$$i\Delta_1 \arg f + i\Delta_2 \arg f + \dots,$$

which is the total variation of the argument of $f(z)$ as z moves around the contour γ. (The point is that we have to keep making possibly different choices of the argument as we go along, depending on the winding behaviour of $f(z)$ around zero.)

Remark 12.2 Given $\gamma : [a, b] \to \mathbb{C}$, let Γ be the contour $\Gamma(t) = f(\gamma(t))$, for $a \le t \le b$. Then

$$\int_\gamma \frac{f'}{f} = \int_a^b \frac{f'(\gamma(t))\gamma'(t)}{f(\gamma(t))} \, dt$$

$$= \int_a^b \frac{\Gamma'(t)}{\Gamma(t)} \, dt$$

$$= \int_\Gamma \frac{dw}{w}$$

$$= 2\pi i \, \text{Ind}(\Gamma : 0).$$

Hence, by the Argument Principle, we see that the number of zeros minus the number of poles inside γ is determined by the winding number of the contour $\Gamma = f \circ \gamma$ about the origin;

$$N_\gamma - P_\gamma = \text{Ind}(\Gamma : 0).$$

In particular, if $f \in H(D)$, then the number of zeros of f inside the closed contour γ is precisely $\text{Ind}(f \circ \gamma : 0)$.

Now for fixed $w_0 \in \mathbb{C}$, any solution to $f(z) = w_0$ is a zero of the function $f(z) - w_0$, and vice versa. By arguing as above (and using $(f - w_0)' = f'$), we see that

$$\int_\gamma \frac{(f - w_0)'}{(f - w_0)} = \int_\gamma \frac{f'}{f - w_0}$$

$$= \int_a^b \frac{f'(\gamma(t))\gamma'(t)}{f(\gamma(t)) - w_0} \, dt$$

$$= \int_\Gamma \frac{dw}{w - w_0}$$

$$= 2\pi i \operatorname{Ind}(\Gamma : w_0)$$

which is to say that the number of solutions to $f(z) = w_0$ inside γ is equal to the winding number $\operatorname{Ind}(f \circ \gamma : w_0)$.

12.3 Rouché's Theorem

This relationship between the number of zeros of a function and certain winding numbers suggests that the comparison of two functions might be attacked by comparing winding numbers. Suppose that γ_1 and γ_2 are two given closed contours, parameterized by $[a, b]$ and neither passing through the origin. If $\gamma_2(t)$ is "close to" $\gamma_1(t)$ for every $t \in [a, b]$, then we would expect that they have the same winding number around 0. We can think of t as "time" and the pair of points $\gamma_1(t)$ and $\gamma_2(t)$ moving as a composite system around 0 as if joined by a spring. An alternative picture is an earth moon system. Here the earth and the moon have the same winding number around the sun, even though they have quite different trajectories.

The issue is what is meant by "close to" in this context? To be sure that the two points $\gamma_1(t)$ and $\gamma_2(t)$ encircle the origin the same number of times, we must ensure that one of them, say $\gamma_2(t)$, cannot "duck under" the origin whilst the other, $\gamma_1(t)$, goes "over the top".

This is ensured if $\gamma_2(t)$ always lies in the disc $D(\gamma_1(t), |\gamma_1(t)|)$, with its centre at $\gamma_1(t)$ and with radius $|\gamma_1(t)|$, since in this case $\gamma_2(t)$ and $\gamma_1(t)$ are always "to the same side of 0". The requirement that $\gamma_2(t)$ belong to $D(\gamma_1(t), |\gamma_1(t)|)$ is to demand that $|\gamma_1(t) - \gamma_2(t)| < |\gamma_1(t)|$.

This discussion leads to the following proposition.

Proposition 12.1 *Suppose γ_1 and γ_2 are closed contours, parameterized by $[a, b]$, such that*

$$|\gamma_1(t) - \gamma_2(t)| < |\gamma_1(t)|$$

for all $a \leq t \leq b$. Then

$$\text{Ind}(\gamma_1 : 0) = \text{Ind}(\gamma_2 : 0).$$

Proof. First we note that neither of γ_1 nor γ_2 pass through 0 (otherwise the inequality above would fail). For $t \in [a, b]$, set $\gamma(t) = \gamma_2(t)/\gamma_1(t)$. Then

$$\frac{\gamma'}{\gamma} = \frac{\gamma_2'}{\gamma_2} - \frac{\gamma_1'}{\gamma_1}$$

and γ is a closed contour satisfying

$$|1 - \gamma(t)| < 1$$

for each $t \in [a, b]$. It follows that $\text{tr}\,\gamma \subset D(1, 1)$ and so, by theorem 8.4 (or, indeed, by theorem 8.2),

$$\text{Ind}(\gamma : 0) = \frac{1}{2\pi i} \int_\gamma \frac{dz}{z} = 0.$$

However,

$$\text{Ind}(\gamma : 0) = \frac{1}{2\pi i} \int_\gamma \frac{dz}{z} = \frac{1}{2\pi i} \int_a^b \frac{\gamma'(t)}{\gamma(t)}\, dt$$

$$= \frac{1}{2\pi i} \int_a^b \frac{\gamma_2'(t)}{\gamma_2(t)}\, dt - \frac{1}{2\pi i} \int_a^b \frac{\gamma_1'(t)}{\gamma_1(t)}\, dt$$

$$= \text{Ind}(\gamma_2 : 0) - \text{Ind}(\gamma_1 : 0),$$

and we conclude that $\text{Ind}(\gamma_1 : 0) = \text{Ind}(\gamma_2 : 0)$. $\qquad\square$

Theorem 12.4 (Rouché's Theorem) *Suppose that f and g are analytic in the star-domain D and that γ is a contour in D, as in Theorem 12.3. Suppose, further, that $|f(\zeta) - g(\zeta)| < |f(\zeta)|$, for all $\zeta \in \text{tr}\,\gamma$. Then f and g have the same number of zeros inside γ (counted according to multiplicity).*

Proof. Let $\gamma_1 = f \circ \gamma$ and $\gamma_2 = g \circ \gamma$ so that $|\gamma_1 - \gamma_2| < |\gamma_1|$. Then, by the Argument Principle, theorem 12.3, and proposition 12.1, we have

$$N_\gamma(f) = \text{Ind}(\gamma_1 : 0) = \text{Ind}(\gamma_2 : 0) = N_\gamma(g)$$

as required. $\qquad\square$

Example 12.1 We shall use Rouché's Theorem to show that $z^5 + 14z + 2$ has precisely 4 zeros inside the annulus $\{ z : \frac{3}{2} < |z| < 2 \}$.

To show this, set $f(z) = z^5$ and $g(z) = z^5 + 14z + 2$. Then $|z| = 2$ implies that $|f(z)| = 32$ and $|f(z) - g(z)| = |14z + 2| \leq 14\,|z| + 2 = 30$. Hence

$|f(z) - g(z)| < |f(z)|$ on the circle $|z| = 2$. By Rouché's Theorem, f and g have the same number of zeros inside the circle, that is, in $\{\, z : |z| < 2 \,\}$. This number is 5. Therefore $z^5 + 14z + 2$ has 5 zeros in $\{\, z : |z| < 2 \,\}$.

Now put $f(z) = 14z$ and leave g as before. For $|z| = \frac{3}{2}$, we find

$$|f(z)| = 14\,|z| = 14 \times \tfrac{3}{2} = 21$$

and

$$|f(z) - g(z)| = \left| z^5 + 2 \right| \leq |z|^5 + 2 = \left(\tfrac{3}{2} \right)^5 + 2 < 10.$$

Therefore $|f(z) - g(z)| < |f(z)|$ on $|z| = \frac{3}{2}$ and so neither f nor g can vanish for such z and $14z$ and $z^5 + 14z + 2$ have the same number of zeros in $\{\, z : |z| < \frac{3}{2} \,\}$, namely 1. It follows that four of the five zeros of the polynomial $z^5 + 14z + 2$ lie in the annulus $\{\, z : \frac{3}{2} < |z| < 2 \,\}$.

As another example, reconsider the Fundamental Theorem of Algebra.

Theorem 12.5 (Fundamental Theorem of Algebra) *For any $n \in \mathbb{N}$ and any complex numbers $a_0, a_1, \ldots, a_{n-1}$, the polynomial*

$$g(z) = z^n + a_{n-1} z^{n-1} + \cdots + a_1 z + a_0$$

has precisely n zeros in the complex plane (including multiplicity).

Proof. Set $f(z) = z^n$. Let $r > |a_0| + \cdots + |a_{n-1}|$ and also $r > 1$. Then, for any z with $|z| = r$, we have

$$
\begin{aligned}
|f(z) - g(z)| &= \left| \sum_{k=0}^{n-1} a_k z^k \right| \\
&\leq \sum_{k=0}^{n-1} |a_k|\,|z^k| \\
&= \sum_{k=0}^{n-1} |a_k|\, r^k \\
&\leq r^{n-1} \sum_{k=0}^{n-1} |a_k|\,, \qquad \text{because } r^k \leq r^{n-1}, \text{ for } 0 \leq k \leq n - 1, \\
&< r^n \\
&= |f(z)|\,.
\end{aligned}
$$

By Rouché's Theorem, it follows that f and g have the same number of zeros inside the circle $\gamma(t) = re^{2\pi i t}$, $0 \leq t \leq 1$, namely n. $\qquad \square$

The following result is a corollary to Rouché's Theorem.

Theorem 12.6   *Suppose that f is analytic in the star-domain D and γ is a contour in D as in Theorem 12.3. Suppose, further, that $f - w_0$ does not vanish on* tr γ. *Then there is $\delta > 0$ such that the equations $w = f(z)$ and $w_0 = f(z)$ have the same number of roots inside γ whenever $w \in D(w_0, \delta)$.*

Proof.   Since $f - w_0$ is continuous and never zero on tr γ and tr γ is compact, it follows that there is some $\delta > 0$ such that $|f(z) - w_0| \geq \delta$ for all $z \in$ tr γ. ($1/(f - w_0)$ is continuous and bounded.)

Let $F(z) = f(z) - w_0$ and $G(z) = f(z) - w$ and suppose $|w - w_0| < \delta$. Then

$$|F(z) - G(z)| = |w - w_0|$$
$$< \delta$$
$$\leq |F(z)|$$

for all $z \in$ tr γ. By Rouché's Theorem, Theorem 12.4, it follows that F and G have the same number of zeros inside γ and the result follows.   □

This has the following interesting corollary.

Theorem 12.7   *Suppose that f is analytic in a domain D and $z_0 \in D$ is such that $f'(z_0) \neq 0$. Then there is some $r > 0$ such that $f : D(z_0, r) \to \mathbb{C}$ is one-one.*

Proof. f has the Taylor expansion $f(z) = \sum_{n=0}^{\infty} a_n (z - z_0)^n$ about z_0, valid in some disc $D(z_0, R)$. Since $f'(z_0) \neq 0$, it follows that $a_1 \neq 0$. Rewrite this expansion as

$$f(z) = a_0 + a_1(z - z_0) + (z - z_0)g(z)$$

where $g(z) = a_2(z - z_0) + a_3(z - z_0)^2 + \dots$ and note that $f(z_0) = a_0$. Since $g(z) \to 0$ as $z \to z_0$, there is $\rho > 0$ such that $|g(z)| < |a_1|$ whenever $z \in \overline{D(z_0, \rho)}$. But since the equality $f(z) = a_0$ entails $(z - z_0)(a_1 + g(z)) = 0$, it follows that $f(z) - a_0$ can have no zeros in the disc $\overline{D(z_0, \rho)}$ apart from z_0 which is a zero of order one. By Theorem 12.6, there is $\delta > 0$ such that if $w \in D(a_0, \delta)$, then $f(z) - w$ also has exactly one zero inside the circle $|z - z_0| = \rho$.

Now let $0 < r < \rho$ be sufficiently small that $|f(z) - f(z_0)| < \delta$ whenever $|z - z_0| < r$. Suppose that $z_1 \in D(z_0, r)$. Then $w = f(z_1) \in D(a_0, \delta)$ and so $f(z) - f(z_1)$ has just one zero inside the circle $|z - z_0| = \rho$, which must therefore be z_1. In other words, $f : D(z_0, r) \to \mathbb{C}$ is one-one.   □

12.4   Open Mapping Theorem

We show that non-constant analytic functions map open sets into open sets. (The image of a constant function is just a single point.)

Theorem 12.8 (Open Mapping Theorem)   *Suppose f is analytic and not constant in a domain D. Then $f(D)$ is an open set. In particular, if $G \subseteq D$ is open, then $f(G)$ is open.*

Proof.   Let $w_0 \in f(D)$. Then there is some $z_0 \in D$ such that $f(z_0) = w_0$. Since f is not constant, the point z_0 is an isolated zero of $f - w_0$ which means that there is $r > 0$ such that $D(z_0, r) \subseteq D$ and $f(z) - w_0$ has no zeros in the punctured disc $D'(z_0, r)$. In particular, $f - w_0$ does not vanish on the circle $|z - z_0| = \frac{1}{2}r$. Let γ be the contour given by $\gamma(t) = \frac{1}{2}re^{2\pi it}$, for $0 \le t \le 1$. By theorem 12.6, we conclude that there is some $\delta > 0$ such that $f(z) - w$ certainly has zeros inside γ whenever $w \in D(w_0, \delta)$. But this simply means that $D(w_0, \delta) \subseteq f(D)$ which shows that $f(D)$ is open.

Suppose $G \subseteq D$ is open. Then G is a union of open discs in D. By the Identity Theorem, f is not constant in any of these discs and so by the first part, the image of any such disc under f is open. But then $f(G)$ is a union of open sets and so is open. $\square$

Chapter 13

Maximum Modulus Principle

13.1 Mean Value Property

The first result of this chapter is a certain mean value property enjoyed by analytic functions and is a direct consequence of Cauchy's Integral Formula.

Theorem 13.1 *Suppose that f is analytic in the disc $D(z_0, R)$. Then, for any $0 < r < R$,*

$$f(z_0) = \frac{1}{2\pi} \int_0^{2\pi} f(z_0 + re^{i\theta}) \, d\theta.$$

Proof. For given $0 < r < R$, let γ be the circle around z_0 with radius r; $\gamma(t) = z_0 + re^{it}$, for $0 \le t \le 2\pi$. Cauchy's Integral Formula, theorem 8.5, gives

$$
\begin{aligned}
f(z_0) &= \frac{1}{2\pi i} \int_\gamma \frac{f(w)}{w - z_0} \, dw \\
&= \frac{1}{2\pi i} \int_0^{2\pi} \frac{f(\gamma(t))}{\gamma(t) - z_0} \, \gamma'(t) \, dt \\
&= \frac{1}{2\pi i} \int_0^{2\pi} \frac{f(\gamma(t))}{re^{it}} \, rie^{it} \, dt \\
&= \frac{1}{2\pi} \int_0^{2\pi} f(z_0 + re^{it}) \, dt
\end{aligned}
$$

and the proof is complete. □

Remark 13.1   Writing $z_0 = x_0 + iy_0$ and $f = u + iv$, we obtain

$$u(x_0, y_0) + iv(x_0, y_0) = \frac{1}{2\pi} \int_0^{2\pi} \big\{ u(x_0 + r\cos\theta, y_0 + r\sin\theta)$$
$$+ iv(x_0 + r\cos\theta, y_0 + r\sin\theta) \big\} \, d\theta.$$

Equating real and imaginary parts gives

$$u(x_0, y_0) = \frac{1}{2\pi} \int_0^{2\pi} u(x_0 + r\cos\theta, y_0 + r\sin\theta) \, d\theta$$

$$v(x_0, y_0) = \frac{1}{2\pi} \int_0^{2\pi} v(x_0 + r\cos\theta, y_0 + r\sin\theta) \, d\theta.$$

For the next result, we recall that a continuous real-valued function on a closed interval $[a, b]$ is bounded and attains its supremum.

Lemma 13.1   *Suppose that $\varphi : [a, b] \to \mathbb{R}$ is continuous and that*

$$\int_a^b \varphi(s) \, ds = M \, (b - a)$$

where $M = \max\{ \varphi(s) : a \le s \le b \}$. Then $\varphi(s) = M$ for all $a \le s \le b$.

Proof.   For $s \in [a, b]$, let $g(s) = M - \varphi(s)$. Then, by hypothesis,

$$\int_a^b g(s) \, ds = M(b - a) - \int_a^b \varphi(s) \, ds = 0.$$

However, $g : [a, b] \to \mathbb{R}$ is continuous and non-negative and so must vanish, i.e., $\varphi = M$ on $[a, b]$. $\square$

13.2   Maximum Modulus Principle

This next major result says that analytic functions do not have maxima, that is, their moduli do not.

Theorem 13.2 (Maximum Modulus Principle)   *Let D be a domain and suppose that $f \in H(D)$. Suppose, further, that there is $M > 0$ such that $|f(z)| \le M$ for all $z \in D$. Then either f is constant on the domain D, or else $|f(z)| < M$ for all $z \in D$. In other words, $|f|$ cannot attain a maximum on D, unless f is constant.*

Proof.   Suppose that $M > 0$ and that $|f(z)| \le M$ for all $z \in D$, and suppose that there is some point $z_0 = x_0 + iy_0 \in D$ such that $|f(z_0)| = M$.

In particular, $|f(z_0)| \neq 0$ and so $f(z)/f(z_0) \in H(D)$ and $|f(z)/f(z_0)| \leq 1$ on the domain D.

Now, $z_0 \in D$ and so there is some $R > 0$ such that $D(z_0, R) \subseteq D$. Let $0 < r < R$. By theorem 13.1,

$$1 = \frac{f(z_0)}{f(z_0)} = \frac{1}{2\pi} \int_0^{2\pi} \frac{f(z_0 + re^{it})}{f(z_0)} \, dt.$$

Writing $f(x+iy)/f(z_0) = u(x,y)+iv(x,y)$, and equating real and imaginary parts, we obtain

$$1 = \frac{1}{2\pi} \int_0^{2\pi} u(x_0 + r\cos t, y_0 + r\sin t) \, dt \qquad (*)$$

and

$$0 = \frac{1}{2\pi} \int_0^{2\pi} v(x_0 + r\cos t, y_0 + r\sin t) \, dt.$$

Let $\varphi(t) = u(x_0 + r\cos t, y_0 + r\sin t)$. Then $\varphi : [0, 2\pi] \to \mathbb{R}$ is continuous and satisfies

$$\varphi(t) = \mathrm{Re}\left(\frac{f(z_0 + re^{it})}{f(z_0)}\right)$$

$$\leq \left|\frac{f(z_0 + re^{it})}{f(z_0)}\right|, \quad \text{using the inequality } \mathrm{Re}\, w \leq |w|,$$

$$= \frac{|f(z_0 + re^{it})|}{M}$$

$$\leq 1.$$

By lemma 13.1, together with the equality $(*)$, it follows that $\varphi(t) = 1$ for all $0 \leq t \leq 2\pi$. Hence

$$1 = |\varphi(t)|^2 = u(x_0 + r\cos t, y_0 + r\sin t)^2$$

$$\leq u(x_0 + r\cos t, y_0 + r\sin t)^2 + v(x_0 + r\cos t, y_0 + r\sin t)^2$$

$$= \left|\frac{f(z_0 + re^{it})}{f(z_0)}\right|^2$$

$$\leq 1.$$

From this, we deduce that $v(x_0 + r\cos t, y_0 + r\sin t) = 0$ and therefore

$$\frac{f(z_0 + re^{it})}{f(z_0)} = 1 + i0,$$

giving $f(z_0 + re^{it}) = f(z_0)$ for all $0 \leq t \leq 2\pi$ and $0 < r < R$. In other words, we have shown that $f(z) = f(z_0)$ for all $z \in D(z_0, R)$. By the Identity Theorem, theorem 8.12, it follows that $f(z) = f(z_0)$ for all z in D.

Thus, if $|f|$ has a maximum in D, then f must be constant in D. If f is non-constant, $|f|$ does not achieve a maximum in D. $\square$

From this, it readily follows that (the modulus of) non-constant analytic functions have no local maxima.

Corollary 13.1   *Suppose that f is analytic on a domain D. Then $|f|$ has no local maxima in D, unless f is constant on D.*

Proof.   Suppose that $z_0 \in D$ is a local maximum for $|f|$, that is, there is some disc $D(z_0, R)$ in D such that $|f(z)| \leq |f(z_0)|$ for all $z \in D(z_0, R)$. If $|f(z_0)| = 0$, then f vanishes on $D(z_0, R)$. On the other hand, suppose that $|f(z_0)| > 0$. Then, by the Maximum Modulus Principle applied to f on the domain $D(z_0, R)$, we find that f is constant on $D(z_0, R)$. In any event, f is constant on the domain D, by the Identity Theorem. $\square$

An alternative version of this can be given using Rouchés Theorem via Theorem 12.8, as follows.

Theorem 13.3   *Suppose f is analytic and non-constant in a domain D. Then for any $z_0 \in D$, there is $z \in D$ such that $|f(z)| > |f(z_0)|$.*

Proof.   Since $f(z_0) \in f(D)$ and $f(D)$ is open (by Theorem 12.8), there is some $\rho > 0$ such that $D(f(z_0), \rho) \subseteq f(D)$. Suppose that $f(z_0) = Re^{i\theta}$ where $R = |f(z_0)|$ and $\theta \in \mathbb{R}$.

Then for any $0 < r < \rho$, $f(z_0) + re^{i\theta} \in D(f(z_0), \rho) \subseteq f(D)$. Hence there is $z \in D$ such that $f(z) = f(z_0) + re^{i\theta}$. But then

$$|f(z)| = \left| Re^{i\theta} + re^{i\theta} \right| = R + r = |f(z_0)| + r > |f(z_0)| ,$$

as required. $\square$

This discussion suggests that if $|f|$ is to achieve a maximum, then this should occur on the boundary of a domain—assuming that the function f is defined there and sufficiently well-behaved. A formulation of this is contained in the next theorem. For this, we note that the closure of a set is the union of the set together with its boundary.

Theorem 13.4   *Let D be a bounded domain and suppose that $f : \overline{D} \to \mathbb{C}$ is continuous and that f is analytic in D. Then either f is constant on D*

or $|f|$ *attains its maximum on the boundary of* D *but not in* D. *In fact, for any* $z \in D$ *(and assuming that* f *is not constant),*

$$|f(z)| < \sup_{w \in D} |f(w)|$$
$$= \max_{\zeta \in \overline{D}} |f(\zeta)| = \max_{\zeta \in \partial D} |f(\zeta)|.$$

Proof. First we note that the boundedness of D implies that of $\overline{D}$. This set is also closed and therefore is compact. The continuity of f on $\overline{D}$ implies that of $|f|$ which therefore is bounded on $\overline{D}$ and achieves its supremum. That is, there is some $\zeta \in \overline{D}$ such that

$$|f(\zeta)| = \sup_{z \in \overline{D}} |f(z)| = M, \text{ say.}$$

By the Maximum Modulus Principle, theorem 13.2, if f is non-constant then $|f(z)| < M$ for all $z \in D$. It follows that $\zeta \in \overline{D} \setminus D = \partial D$, the boundary of the domain D.   □

Remark 13.2 A slight rephrasing of this theorem is as follows. Suppose that D is a domain with $\overline{D}$ bounded and suppose that $f : \overline{D} \to \mathbb{C}$ is continuous and that f is analytic in D. If $|f(z)| \leq M$ for all $\zeta \in \partial D$, then either $|f(z)| < M$ for all $z \in D$, or else f is constant on D.

Example 13.1 For unbounded regions the above result may be false. For example, let D be the infinite horizontal strip

$$D = \{ z : -\frac{\pi}{2} < \text{Im}\, z < \frac{\pi}{2} \},$$

and let $f(z) = \exp(\exp z)$, $z \in \mathbb{C}$. Then f is entire and so is certainly continuous on $\overline{D}$.

We claim that f is bounded on the boundary of D. To see this, let $\zeta \in \partial D$, so that $\zeta = x \pm i\frac{\pi}{2}$ for some $x \in \mathbb{R}$. We have

$$f(\zeta) = \exp(\exp(x \pm i\pi/2))$$
$$= \exp(e^x \, e^{\pm i\pi/2})$$
$$= \exp(\pm i e^x)$$
$$= \cos e^x \pm i \sin e^x.$$

It follows that $|f(\zeta)| = 1$ for every $\zeta \in \partial D$, the boundary of D.

Is $|f(z)| < 1$ for all $z \in D$? The answer is no. For example, suppose that $z = x \in \mathbb{R} \subseteq D$. Then we find that

$$f(z) = f(x) = \exp(\exp x) = e^{e^x}.$$

Clearly $|f(x)| = e^{e^x} \to \infty$ as $x \to \infty$ and so $|f|$ is not even bounded on D never mind being less than 1.

13.3   Minimum Modulus Principle

Maxima of $|f|$ correspond to minima of $|\frac{1}{f}|$, assuming that we do not need to worry about f being zero. This observation leads to the following Minimum Modulus Principle.

Theorem 13.5 (Minimum Modulus Principle)   *Suppose f is analytic and non-constant on a domain D and that $|f(z)| \geq m > 0$, for all $z \in D$. Then $|f(z)| > m$ for all $z \in D$.*

If, in addition, D is bounded and f is defined and continuous and non-zero on $\overline{D}$, then

$$|f(z)| > \inf_{w \in D} |f(w)| = \min_{\zeta \in \partial D} |f(\zeta)|$$

for all $z \in D$.

Proof.   Apply the Maximum Modulus Principle to $g = \dfrac{1}{f}$.   □

As motivation for the next theorem, we observe that if $w = a + ib$, then $e^w = e^a e^{ib}$, so that $|e^w| = e^a$. This idea of looking at exponentials leads to maximum and minimum principles for the real and imaginary parts of an analytic function.

Theorem 13.6   *Let D be a domain and suppose that $f = u + iv \in H(D)$ is non-constant. Suppose M is some real number such that $u \leq M$ on D. Then $u < M$ on D. Similarly, if there is $m \in \mathbb{R}$ such that $m \leq u$ on D, then $m < u$ on D. A similar pair of statements hold for v.*

Furthermore, if D is bounded and f is defined and continuous on $\overline{D}$, then the suprema and infima of u and v over $\overline{D}$ are attained on the boundary of the domain D.

Proof.   Suppose that $u(x, y) \leq M$ for $z = x + iy \in D$. Set

$$g(z) = \exp(f(z)) = e^u \, e^{iv}, \text{ for } z \in D.$$

Then g is non-constant on D (otherwise f would be) and $|g(z)| = e^{u(x,y)}$, so that $u \leq M$ implies that $e^u \leq e^M$. (The real exponential function is monotonic increasing.) Hence $|g(z)| \leq e^M$ on D. By the Maximum Modulus Principle, theorem 13.2, it follows that

$$e^u = |g(z)| < e^M$$

and therefore $u(x,y) < M$ for all $x + iy \in D$.

Next, suppose that $m \leq u(x,y)$ for all $z = x + iy \in D$. Put

$$h(z) = \exp(-f(z)) = e^{-u} e^{-iv}, \text{ for } z \in D.$$

Then, h is non-constant on D and, for any $z \in D$,

$$|h(z)| = e^{-u} \leq e^{-m}$$

so that

$$e^{-u} = |h(z)| < e^{-m}$$

on D, by the Maximum Modulus Principle. It follows that $m < u(x,y)$ for any $x + iy \in D$, as required. The analogous results for v are obtained by considering the functions $\exp(\mp i f(z))$.

Finally, we note that g and h are defined and continuous on $\overline{D}$ if f is. Furthermore, if f is non-constant, neither are g nor h. By the Maximum Modulus Principle, the suprema of the moduli of these functions is attained on the boundary of D (and not on D itself). This amounts to saying that the maximum and the minimum of u over $\overline{D}$ are both attained on the boundary of D (and not in D). A similar argument applied to the two functions $\exp(\mp i f)$ leads to the similar statements for v rather than u. $\square$

13.4 Functions on the Unit Disc

Theorem 13.7 (Schwarz's Lemma) *Suppose that f is analytic in the open unit disc $D(0,1)$ and satisfies $f(0) = 0$ and $|f(z)| \leq M$ for all $|z| < 1$. Then $|f(z)| \leq M|z|$ for all $|z| < 1$.*

Proof. Let

$$f(z) = a_0 + a_1 z + a_2 z^2 + \ldots$$

be the Taylor series expansion of $f(z)$ about $z_0 = 0$, for $z \in D(0,1)$. Then $a_0 = f(0) = 0$, by hypothesis.

Put $g(z) = a_1 + a_2 z + \ldots$, so that $f(z) = z\, g(z)$. The function g is analytic in $D(0,1)$ and $g(z) = f(z)/z$ for $0 < |z| < 1$.

Let $0 < r < 1$ and suppose that $|z| = r$. Then

$$|g(z)| = \left| \frac{f(z)}{z} \right| = \frac{|f(z)|}{r} \le \frac{M}{r}\,.$$

By the Maximum Modulus Principle, it follows that

$$|g(z)| \le \frac{M}{r}$$

for all $|z| \le r$. Letting $r \to 1$ we see that $|g(z)| \le M$, for all $|z| < 1$, and therefore

$$|f(z)| = |z|\,|g(z)| \le M\,|z|$$

for all $z \in D(0,1)$, as required. $\qquad\qquad\qquad\qquad\qquad\qquad\qquad\square$

Remark 13.3   In fact, the inequality $|g(z)| \le M$ for $|z| < 1$ implies that $|g(z)| < M$ for all $|z| < 1$ or else g is a constant, $g(z) = \alpha$ and $|\alpha| = M$. But then

$$|f(z)| = |z\, g(z)| < M\,|z|\,,$$

for all $0 < |z| < 1$, or else $f(z) = \alpha\, z$, with $|\alpha| = M$.

We can use Schwarz's Lemma to classify those mappings of the open unit disc $D(0,1)$ onto itself which are analytic, one-one and with analytic inverse. First we consider such maps which also preserve the origin.

Theorem 13.8   *Suppose $f : D(0,1) \to D(0,1)$ is analytic and satisfies:*

(i) *$f : D(0,1) \to D(0,1)$ is one-one and onto,*

(ii) *$f^{-1} : D(0,1) \to D(0,1)$ is analytic,*

(iii) *$f(0) = 0$.*

Then $f(z) = \alpha z$ for some $\alpha \in \mathbb{C}$ with $|\alpha| = 1$, for all $z \in D(0,1)$.

Proof.   By hypothesis, f is analytic, $f(0) = 0$ and $|f(z)| < 1$ for all z in the disc $D(0,1)$. By Schwarz's Lemma, it follows that $|f(z)| \le |z|$ for all $z \in D(0,1)$. Exactly the same reasoning applied to the inverse function f^{-1} implies that $\left| f^{-1}(z) \right| \le |z|$ for all $z \in D(0,1)$. Hence

$$|z| = \left| f^{-1}(f(z)) \right| \le |f(z)| \le |z|$$

and we see that $|f(z)| = |z|$ for all $z \in D(0,1)$. Let $h(z) = f(z)/z$ for z in the punctured disc $D'(0,1)$. Then h is analytic in this punctured disc and obeys $|h(z)| = 1$ there. We deduce that h is constant on $D'(0,1)$, i.e., $h(z) = \alpha$ for some $\alpha \in \mathbb{C}$ with $|\alpha| = 1$. Thus $f(z) = \alpha z$ for all $z \in D'(0,1)$. This equality persists even for $z = 0$ since $f(0) = 0$ and the proof is complete. $\qquad\square$

Before considering the general case (i.e., removing the assumption that $f(0) = 0$), we need one more observation.

Theorem 13.9 *For any $a \in D(0,1)$, $z \mapsto g_a(z) = \dfrac{z - a}{1 - \overline{a}\,z}$ is a one-one mapping of $D(0,1)$ onto itself.*

Proof. Let $z \in D(0,1)$ and set $w = g_a(z)$. Then

$$
\begin{aligned}
1 - |w|^2 &= 1 - \overline{w}\,w \\
&= 1 - \frac{(\overline{z} - \overline{a})(z - a)}{(1 - a\,\overline{z})(1 - \overline{a}\,z)} \\
&= \frac{(1 - |a|^2)(1 - |z|^2)}{(1 - a\,\overline{z})(1 - \overline{a}\,z)} \\
&= \frac{(1 - |a|^2)(1 - |z|^2)}{|1 - \overline{a}\,z|^2} > 0
\end{aligned}
$$

and so we see that $g_a : D(0,1) \to D(0,1)$.

Furthermore, g_{-a} is the inverse of g_a and so we deduce that g_a is both one-one and onto $D(0,1)$. $\qquad\square$

Theorem 13.10 *Suppose that $f : D(0,1) \to D(0,1)$ is analytic, one-one, onto and such that f^{-1} is analytic. Then*

$$
f(z) = \alpha \left(\frac{z - a}{1 - \overline{a}\,z} \right)
$$

for some $a \in D(0,1)$ and some $\alpha \in \mathbb{C}$ with $|\alpha| = 1$.

Proof. Since f is one-one and onto $D(0,1)$, there is a unique $a \in D(0,1)$ such that $f(a) = 0$. Let g_{-a} be the transformation $z \mapsto (z + a)/(1 + \overline{a}\,z)$, as above, and let φ be the composition $\varphi : z \mapsto f(g_{-a}(z))$. This is a composition of analytic one-one mappings, each mapping the disc $D(0,1)$ onto itself and each with an analytic inverse. The same is therefore true of φ. Furthermore, by construction, $\varphi(0) = 0$. It follows that $\varphi(z) = \alpha z$ for some $\alpha \in \mathbb{C}$ with $|\alpha| = 1$. Hence $\varphi(g_a(z)) = \alpha g_a(z)$, i.e., $f(z) = \alpha g_a(z)$ for all $z \in D(0,1)$. $\qquad\square$

13.5 Hadamard's Theorem and the Three Lines Lemma

The following is a maximum modulus result on an unbounded domain.

Theorem 13.11 (Hadamard) *Suppose that f is analytic in the (vertical) strip $S = \{ z : 0 < \operatorname{Re} z < 1 \}$ and continuous on its closure $\overline{S}$. Suppose, further, that $|f(z)| \leq K$ for all $z \in S$ and that $|f(z)| \leq 1$ for all $z \in \partial S$. Then $|f(z)| \leq 1$ for all $z \in S$.*

Proof. For $n \in \mathbb{N}$, let $g_n(z) = f(z) \exp\left(\frac{z^2-1}{n}\right)$. Then, for each fixed $z \in S$, $g_n(z) \to f(z)$, as $n \to \infty$, since $\exp(\frac{1}{n}(z^2 - 1)) \to 1$, as $n \to \infty$. Moreover, with $z = x + iy$,

$$\left| \exp\frac{(z^2-1)}{n} \right| = \left| \exp\left(\frac{x^2 - y^2 - 1}{n} + \frac{2ixy}{n}\right) \right|$$

$$= \exp\left(\frac{x^2 - y^2 - 1}{n}\right)$$

$$\leq \exp\left(-\frac{y^2}{n}\right), \quad \text{since } x^2 \leq 1 \text{ for } z \in \overline{S}.$$

Fix $z \in S$ and $n \in \mathbb{N}$. Choose $y_0 > |\operatorname{Im} z|$ so large that $K e^{-y_0^2/n} < 1$ and let R be the rectangle

$$R = \{ z : 0 < \operatorname{Re} z < 1, -y_0 < \operatorname{Im} z < y_0 \}.$$

Then z belongs to the interior of R.

We apply the Maximum Modulus Principle to the function g_n on R to obtain

$$|g_n(z)| \leq \max_{\zeta \in \partial R} |g_n(\zeta)| \leq 1.$$

Therefore $|f(z)| = \lim_{n \to \infty} |g_n(z)| \leq 1$ for any $z \in S$. $\qquad \square$

Corollary 13.2 (Three Lines Lemma) *Suppose that f is as in the theorem, but satisfies the bounds $|f(z)| \leq K$ on S, $|f(z)| \leq M_0$ when $\operatorname{Re} z = 0$, and $|f(z)| \leq M_1$ when $\operatorname{Re} z = 1$. Then*

$$|f(z)| \leq M_0^{1-\operatorname{Re} z} M_1^{\operatorname{Re} z} \quad \text{for all } z \in S.$$

In other words, $|f(x+iy)| \leq M_0^{1-x} M_1^x$ for all $0 \leq x \leq 1$ and $y \in \mathbb{R}$, and so

$$M_x \leq M_0^{1-x} M_1^x$$

where $M_x = \sup_y |f(x+iy)|$.

Proof.   Set $h(z) = f(z) M_0^{z-1} M_1^{-z}$, for $z \in \overline{S}$ (principal values). Now,

$$
\begin{aligned}
|M_0^{z-1}| &= |M_0^{\operatorname{Re} z - 1} M_0^{i \operatorname{Im} z}| \\
&= |\exp((\operatorname{Re} z - 1) \operatorname{Log} M_0) \exp(i \operatorname{Im} z \operatorname{Log} M_0)| \\
&= \exp((\operatorname{Re} z - 1) \operatorname{Log} M_0) \\
&= M_0^{\operatorname{Re} z - 1}.
\end{aligned}
$$

So

$$
|M_0^{z-1}| = M_0^{\operatorname{Re} z - 1} = \begin{cases} M_0^{-1}, & \text{if } \operatorname{Re} z = 0 \\ 1, & \text{if } \operatorname{Re} z = 1. \end{cases}
$$

Similarly,

$$
|M_1^{-z}| = M_1^{-\operatorname{Re} z} = \begin{cases} 1, & \text{if } \operatorname{Re} z = 0 \\ M_1^{-1}, & \text{if } \operatorname{Re} z = 1. \end{cases}
$$

Hence $|h| \leq 1$ on ∂S, and, by the theorem, $|h(x+iy)| \leq 1$ on S. That is,

$$
|f(z)| \, |M_0^{z-1}| \, |M_1^{-z}| \leq 1
$$

on S, or

$$
|f(z)| \leq M_0^{1 - \operatorname{Re} z} M_1^{\operatorname{Re} z}
$$

on S, as required.

This, together with the hypotheses, means that

$$
|f(x+iy)| \leq M_0^{1-x} M_1^{x}
$$

for all $0 \leq x \leq 1$ and $y \in \mathbb{R}$. Since the right hand side is an upper bound for the left hand side, as y varies in $\mathbb{R}$, we conclude that

$$
\sup_{y \in \mathbb{R}} |f(x+iy)| \leq M_0^{1-x} M_1^{x},
$$

as claimed. $\square$

Chapter 14

Möbius Transformations

14.1 Special Transformations

We can think of a complex function $f : \mathbb{C} \to \mathbb{C}$ as a transformation of the complex plane into itself. This introduces a strong geometrical flavour to the discussion.

We first consider the four basic transformations of translation, rotation, magnification and so-called inversion. A translation is any transformation of the form $z \mapsto w = z + a$, with $a \in \mathbb{C}$. A rotation is any map of the form $z \mapsto w = e^{i\varphi}z$, with $\varphi \in \mathbb{R}$. A magnification is a map of the form $z \mapsto w = rz$ with $r > 0$. Note that if $r > 1$, then this is a magnification in the conventional sense, but if $r < 1$ then it is a contraction rather than a magnification. Illustrations of these three mappings are given in Figs. 14.1, 14.2 and 14.3, below.

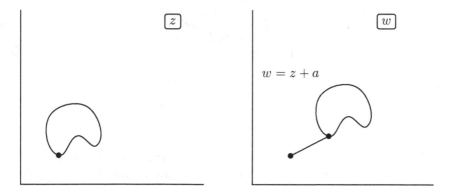

Fig. 14.1 Translation: $z \mapsto w = z + a$, $a \in \mathbb{C}$.

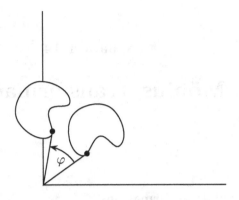

Fig. 14.2   Rotation: $z \mapsto w = e^{i\varphi}z,\ \varphi \in \mathbb{R}$.

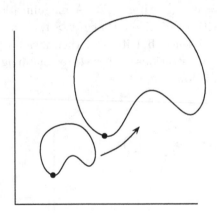

Fig. 14.3   Magnification: $z \mapsto w = rz,\ r > 0$.

Notice that under any of these three special transformations, any given region retains its general shape but may be moved around and magnified (or shrunk).

Next, we discuss inversion, which is somewhat less transparent.

14.2 Inversion

Inversion is the map $z \mapsto w = \dfrac{1}{z}$, for $z \neq 0$. To see what happens under an inversion, write $z = r\, e^{i\theta}$. Then

$$w = \frac{1}{z} = \frac{1}{r} e^{-i\theta}.$$

Thus, points near zero are transformed far away, points in the upper half-plane are mapped into the lower half-plane and vice versa.

Indeed, we can write $z \mapsto w = \dfrac{1}{z}$ as $z \mapsto w = \dfrac{1}{z} = \dfrac{\bar{z}}{|z|^2}$.

Let us see what happens to circles and straight lines under an inversion. To say that z lies on a circle, we mean that if $z = x + iy$, then x and y satisfy an equation of the form

$$a(x^2 + y^2) + bx + cy + d = 0 \qquad (*)$$

for real numbers a, b, c, d, with $a \neq 0$. Explicit inclusion of the coefficient a means that we can set $a = 0$ to get the equation of a straight line. That is, equation $(*)$ determines either a circle (if $a \neq 0$) or a straight line (if $a = 0$) in $\mathbb{C}$. This circle/line passes through the origin depending on whether $d = 0$ or not.

Actually, if $a \neq 0$, then by completing the square, we see that $(*)$ can be written as

$$\left(x + \tfrac{b}{2a}\right)^2 + \left(y + \tfrac{c}{2a}\right)^2 + \tfrac{d}{a} - \tfrac{b^2}{4a^2} - \tfrac{c^2}{4a^2} = 0.$$

This really is a circle (with strictly positive radius) provided $b^2 + c^2 > 4ad$.

Let us write $\dfrac{1}{z} = w = u + iv$. Then

$$z = \frac{1}{w} = \frac{1}{u+iv} = \frac{u-iv}{u^2+v^2} = x + iy.$$

It follows that

$$x = \frac{u}{u^2+v^2} \quad \text{and} \quad y = -\frac{v}{u^2+v^2}.$$

Now, if x and y satisfy $(*)$, then

$$\frac{a(u^2+v^2)}{u^2+v^2} + \frac{bu}{u^2+v^2} - \frac{cv}{u^2+v^2} + d = 0$$

and so

$$a + bu - cv + d(u^2 + v^2) = 0 \qquad (**)$$

i.e., (u, v) lies on a straight line or a circle (depending on whether $d = 0$ or not). It follows that the family of circles and straight lines is mapped into itself under the inversion $z \mapsto w = \dfrac{1}{z}$ (the points $z = 0$ and $w = 0$ excepted).

	$(*)$	$(**)$
$a \neq 0,\ d \neq 0$	circle not through $(0,0)$	circle not through $(0,0)$
$a = 0,\ d \neq 0$	line not through $(0,0)$	circle through $(0,0)$
$a \neq 0,\ d = 0$	circle through $(0,0)$	line not through $(0,0)$
$a = 0,\ d = 0$	line through $(0,0)$	line through $(0,0)$

So the family of circles and straight lines is mapped into itself under all four of the special maps considered so far, and therefore also under any composition of such maps.

14.3 Möbius Transformations

Definition 14.1 A Möbius transformation is a map of the form

$$z \mapsto T(z) = \frac{az + b}{cz + d}$$

where $a, b, c, d \in \mathbb{C}$ satisfy $ad - bc \neq 0$.

Möbius transformations are also called bilinear transformations or fractional transformations. The condition $ad - bc \neq 0$ ensures that the map is defined (except, of course, at $z = -d/c$ if $c \neq 0$) and is not a constant. If $a \neq 0$ and $c \neq 0$, then $T(z) = a(acz + bc)/c(acz + ad)$. If $bc = ad$ then this reduces to a constant, namely a/c. Also, one can check that the condition $ad - bc \neq 0$ means that $T(z) = T(w)$ if and only if $z = w$.

If $w = T(z) = (az + b)/(cz + d)$, then "solving for z", we calculate that

$z = (-dw + b)/(cw - a)$ and $(-d)(-a) - bc = ad - bc \neq 0$. Hence

$$S(w) = \frac{-dw + b}{cw - a}$$

is also a Möbius transformation and is the inverse of T. Furthermore, if T_1 and T_2 are Möbius transformations, then so is their composition $T_2 T_1$ (which maps z into $T_2(T_1(z))$). It follows that the set of Möbius transformations forms a group (under composition) with the group identity being the identity transformation $I(z) = z$.

It turns out that any any Möbius transformation $T(z) = \dfrac{az + b}{cz + d}$ can be expressed in terms of the special transformations considered earlier. To see this, suppose first that $c \neq 0$. Then we write

$$\begin{aligned}
\frac{az + b}{cz + d} &= \frac{(a/c)(cz + d) - (ad/c) + b}{cz + d} \\
&= \frac{a}{c} - \frac{(ad/c) - b}{cz + d} \\
&= \frac{a}{c} + \frac{bc - ad}{c^2(z + (d/c))}.
\end{aligned}$$

Let $\lambda = \dfrac{bc - ad}{c^2} \neq 0$. Then

$$T(z) = \frac{\lambda}{(z + (d/c))} + \frac{a}{c}.$$

Consider the following maps:

$$z \mapsto w_1 = z + \frac{d}{c} \qquad\qquad \text{translation}$$

$$w_1 \mapsto w_2 = \frac{1}{w_1} \qquad\qquad \text{inversion}$$

$$w_2 \mapsto w_3 = |\lambda|\, e^{i\arg \lambda} w_2 \qquad\qquad \text{magnification/rotation}$$

$$w_3 \mapsto w = w_3 + \frac{a}{c} \qquad\qquad \text{translation.}$$

Then

$$w = w_3 + \frac{a}{c} = \lambda w_2 + \frac{a}{c} = \frac{\lambda}{w_1} + \frac{a}{c} = \frac{\lambda}{z + (d/c)} + \frac{a}{c}$$

i.e., $z \mapsto T(z) = w$ is given by the composition

$$z \underset{\text{trans}}{\longmapsto} w_1 \underset{\text{inv}}{\longmapsto} w_2 \underset{\text{mag/rot}}{\longmapsto} w_3 \underset{\text{trans}}{\longmapsto} w.$$

Now consider the case $c = 0$, which is a little simpler. First note that now $ad \neq bc$ demands that $ad \neq 0$. In particular, $d \neq 0$. Therefore

$$T(z) = \frac{az + b}{d} = \frac{a}{d}\, z + \frac{b}{d}\,.$$

Consider

$$z \mapsto w_1 = \frac{a}{d}\, z \qquad\qquad\qquad \text{magnification/rotation}$$

$$w_1 \mapsto w = w_1 + \frac{b}{d} \qquad\qquad\qquad \text{translation}$$

giving $T(z) = w$ via

$$z \underset{\text{mag/rot}}{\longmapsto} w_1 \underset{\text{trans}}{\longmapsto} w.$$

We have seen that any Möbius transformation can be expressed as a suitable composition of the special mappings, namely, translation, rotation, magnification and inversion. It follows that the family of circles and straight lines is mapped into itself under any Möbius transformation.

Let us consider the inversion $z \mapsto 1/z$ in more detail. Consider points $z \in \mathbb{C}$ with $\operatorname{Re} z$ constant, that is points of the form $z = \alpha + iy$, $y \in \mathbb{R}$, for fixed $\alpha \in \mathbb{R}$. Such zs form a line parallel to the imaginary axis. Suppose that $\alpha \neq 0$, so that the line does not pass through the origin.

Now, if $\dfrac{1}{z} = w = u + iv$, then $z = \dfrac{1}{u + iv}$ and so $z = \alpha + iy = \dfrac{u - iv}{u^2 + v^2}$ giving

$$\alpha = \frac{u}{u^2 + v^2} \quad \text{and} \quad y = -\frac{v}{u^2 + v^2}\,.$$

That is, w is such that u and v satisfy $\alpha = \dfrac{u}{u^2 + v^2}$, or $(u^2 + v^2) - \dfrac{u}{\alpha} = 0$ or

$$\left(u - \frac{1}{2\alpha}\right)^2 + v^2 = \frac{1}{4\alpha^2}\,.$$

This is the equation of a circle, centred at $(\frac{1}{2\alpha}, 0)$ and with radius $\frac{1}{2|\alpha|}$. So w lies on this circle. To find the range of values taken by w, we use

$$w = u + iv = \frac{1}{z} = \frac{1}{\alpha + iy} = \frac{\alpha - iy}{\alpha^2 + y^2}$$

to get

$$u = \frac{\alpha}{\alpha^2 + y^2} \quad \text{and} \quad v = -\frac{y}{\alpha^2 + y^2}\,.$$

As z varies on the line, i.e., as y varies in $\mathbb{R}$, we see that u takes all values between 0 and $1/\alpha$, including $1/\alpha$ but not including 0. Thus $(u - \frac{1}{2\alpha})$ takes on all values between $\pm\frac{1}{2\alpha}$, not including $-\frac{1}{2\alpha}$ and v takes on all values between $\pm\frac{1}{2\alpha}$ (including both values $\pm\frac{1}{2\alpha}$ themselves). It follows that, for any $\alpha \neq 0$, the image of the line $\{\, z : \operatorname{Re} z = \alpha \,\}$ under the inversion $z \mapsto w = \dfrac{1}{z}$ is the set $\{\, w : |w - \frac{1}{2\alpha}| = \frac{1}{2|\alpha|} \,\} \setminus \{0\}$. This leads to the following theorem.

Theorem 14.1 *Inversion* $z \mapsto w = \dfrac{1}{z}$ *maps the half-plane* $\{\, z : \operatorname{Re} z > 1 \,\}$ *one-one onto the disc* $\{\, w : |w - \frac{1}{2}| < \frac{1}{2} \,\}$, *and vice versa.*

Proof. We can use the circles, varying $\alpha > 1$, as above, or prove this directly as follows. Setting $z = x + iy$, $w = 1/z$, we have

$$\left| w - \frac{1}{2} \right| = \left| \frac{1}{z} - \frac{1}{2} \right| < \frac{1}{2} \iff \left| \frac{2 - z}{2z} \right| < \frac{1}{2}$$
$$\iff |2 - z| < |z|$$
$$\iff |2 - z|^2 < |z|^2$$
$$\iff (2 - x)^2 + y^2 < x^2 + y^2$$
$$\iff 4 - 4x + x^2 + y^2 < x^2 + y^2$$
$$\iff 4(1 - x) < 0$$
$$\iff x > 1.$$

Hence inversion maps the half-plane $\{\, z : \operatorname{Re} z > 1 \,\}$ into the disc $D(\frac{1}{2}, \frac{1}{2})$ and it also maps the disc $D(\frac{1}{2}, \frac{1}{2})$ into the half-plane $\{\, z : \operatorname{Re} z > 1 \,\}$. But inversion is its own inverse, so it follows that it maps $\{\, z : \operatorname{Re} z > 1 \,\}$ onto $D(\frac{1}{2}, \frac{1}{2})$ and $D(\frac{1}{2}, \frac{1}{2})$ onto $\{\, z : \operatorname{Re} z > 1 \,\}$. $\square$

We can use this fact to help construct Möbius transformations mapping various given half-planes into discs and vice versa by manœuvering via the "standard" half-plane/disc pair, as above.

Example 14.1 We shall find a Möbius transformation which maps the half-plane $\{\, z : \operatorname{Im} z > 2 \,\}$ onto the disc $\{\, w : |w - 2| < 3 \,\}$.

As can be seen from Fig. 14.4, the idea is to manipulate the original half-plane into the standard half-plane $\{\, w_2 : \operatorname{Re} w_2 > 1 \,\}$ using rotations and translations (in fact, one of each). Inversion now transforms this standard half-plane into the standard disc. Further rotations, translations and magnifications now bring the image into the required position.

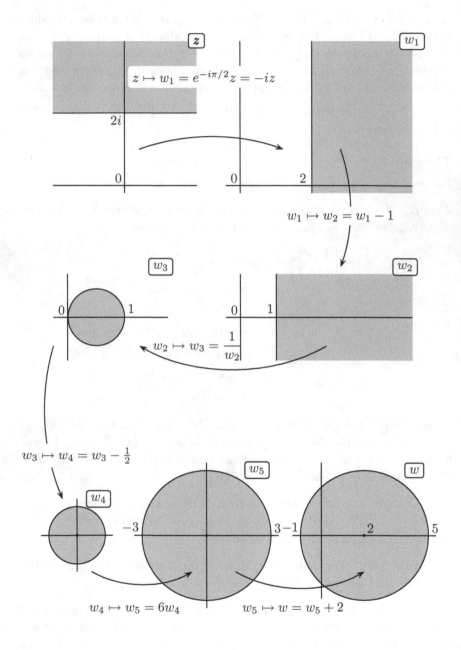

Fig. 14.4   Manœuvring via the standard half-plane/disc pair.

The overall transformation is built from a number of simple ones:

$$z \mapsto w_1 = e^{-i\pi/2}z = -iz, \quad w_1 \mapsto w_2 = w_1 - 1,$$

$$w_2 \mapsto w_3 = \frac{1}{w_2}, \quad w_3 \mapsto w_4 = w_3 - \tfrac{1}{2},$$

$$w_4 \mapsto w_5 = 6\,w_4, \quad w_5 \mapsto w = w_5 + 2.$$

Combining these, we get

$$
\begin{aligned}
w = w_5 + 2 &= 6w_4 + 2 = 6\big(w_3 - \tfrac{1}{2}\big) + 2 \\
&= 6w_3 - 1 = \frac{6}{w_2} - 1 = \frac{6}{w_1 - 1} - 1 \\
&= \frac{6}{-iz - 1} - 1 = \frac{6 + iz + 1}{-iz - 1} \\
&= \frac{iz + 7}{-iz - 1}.
\end{aligned}
$$

14.4 Möbius Transformations in the Extended Complex Plane

The Möbius transformation $T : z \mapsto (az + b)/(cz + d)$ is not defined when $z = -d/c$ (when $c \neq 0$). Now, we can extend the inversion mapping $z \mapsto 1/z$ on $\mathbb{C} \setminus \{0\}$ to a mapping from $\mathbb{C}_\infty$ to $\mathbb{C}_\infty$ by the assignments $0 \mapsto \infty$ and $\infty \mapsto 0$. We mimic this for any Möbius transformation as follows. First recall that $ad - bc \neq 0$, so c and d cannot both be zero. If $c = 0$ (so that $a \neq 0$), set

$$
Tz = \begin{cases} (az + b)/d, & \text{for } z \in \mathbb{C}, \\ \infty, & z = \infty. \end{cases}
$$

If $c \neq 0$, set

$$
Tz = \begin{cases} (az + b)/(cz + d), & \text{for } z \in \mathbb{C} \text{ and } z \neq -d/c, \\ \infty, & z = -d/c \\ a/c, & z = \infty. \end{cases}
$$

In this way, any Möbius transformation $T : z \mapsto (az + b)/(cz + d)$ can be extended to a mapping on $\mathbb{C}_\infty$. Moreover, one checks that the mapping $z \mapsto (dz - b)/(-cz + a)$ is the inverse to T on $\mathbb{C}_\infty$, so that the collection of Möbius transformations on $\mathbb{C}_\infty$ also forms a group under composition.

Proposition 14.1 *Let z_1, z_2 and z_3 be any three distinct points in $\mathbb{C}_\infty$. Then there is a Möbius transformation mapping the triple (z_1, z_2, z_3) to $(0, 1, \infty)$.*

Proof. Suppose first that z_1, z_2 and z_3 all belong to $\mathbb{C}$. Set

$$Tz = \frac{(z - z_1)(z_2 - z_3)}{(z_2 - z_1)(z - z_3)}.$$

Then we see that T is a Möbius transformation with the required properties.

Now, the cases when one of z_1, z_2 or z_3 is equal to ∞ are handled as follows. We define T by

$$Tz = \begin{cases} (z_2 - z_3)/(z - z_3), & \text{if } z_1 = \infty, \\ (z - z_1)/(z - z_3), & \text{if } z_2 = \infty, \\ (z - z_1)/(z_2 - z_3), & \text{if } z_3 = \infty. \end{cases}$$

Again, one sees that T has the required properties. □

Theorem 14.2 *Suppose that z_1, z_2 and z_3 are any distinct points in $\mathbb{C}_\infty$ and that S is a Möbius transformation such that $Sz_j = z_j$ for $j = 1, 2, 3$. Then $Sz = z$ for all $z \in \mathbb{C}_\infty$.*

Proof. Suppose first that $(z_1, z_2, z_3) = (0, 1, \infty)$. Writing $Sz = \dfrac{az + b}{cz + d}$, we calculate that

$$0 \mapsto 0 \implies 0 = b/d \implies b = 0$$

and

$$1 \mapsto 1 \text{ and } \infty \mapsto \infty \implies a/(c + d) = 1 \text{ and } c = 0 \implies a = d.$$

It follows that $Sz = z$ for all z.

Now consider the general case of any distinct points z_1, z_2 and z_3. Let T be a Möbius transformation which maps (z_1, z_2, z_3) to $(0, 1, \infty)$. Then the composition $T \circ S \circ T^{-1}$ is a Möbius transformation mapping $(0, 1, \infty)$ to $(0, 1, \infty)$ and so according to the above argument it is equal to the identity transformation. It follows that $S = T^{-1} \circ T$, that is, S is also the identity transformation, $Sz = z$ for all z. □

Corollary 14.1   *If S and T are Möbius transformations obeying $Sz_j = Tz_j$ for any three distinct points z_1, z_2 and z_3 in $\mathbb{C}_\infty$, then $S = T$.*

Proof.   We simply note that each z_j, $j = 1, 2, 3$, is fixed under the map $S \circ T^{-1}$ so that $S \circ T^{-1}z = z$ for all z, by the theorem. It follows that $Sz = Tz$ for all z. $\square$

Chapter 15

Harmonic Functions

15.1 Harmonic Functions

We recall that if $f = u + iv$ is analytic in a domain D then it obeys the Cauchy-Riemann equations there; $u_x = v_y$ and $u_y = -v_x$. Indeed, we have seen that

$$f' = u_x + iv_x = v_y - iu_y \, .$$

Moreover, we know that f has derivatives of any order and so u and v also possess partial derivatives of any order. It follows that $v_{xy} = v_{yx}$ and so we find that $u_{xx} = (u_x)_x = (v_y)_x = (v_x)_y = -u_{yy}$, that is, u satisfies Laplace's equation

$$u_{xx} + u_{yy} = 0$$

in the domain D.

Definition 15.1 We say that the function $\varphi(x,y)$ is harmonic in the domain D in $\mathbb{R}^2$ if all partial derivatives of φ up to second order exist and are continuous in D and if φ satisfies Laplace's equation $\varphi_{xx} + \varphi_{yy} = 0$ in D.

We can summarize the remarks above by saying that the real part of a function analytic in a domain D is harmonic there. Note also that if φ is harmonic in D, then the mixed partial derivatives φ_{xy} and φ_{yx} are equal in D.

It is natural to ask whether every harmonic function is the real part of some analytic function.

Example 15.1 Let $\varphi(x,y) = \log \sqrt{x^2 + y^2}$ for $(x,y) \neq (0,0)$. Then one checks that φ is harmonic in this region. However, $\varphi(x,y) = \operatorname{Re} \operatorname{Log} z$

219

where $z = x + iy$ and it follows that φ is not the real part of any function analytic in the punctured plane $D = \mathbb{C} \setminus \{0\}$. In fact, if $f \in H(D)$ and if $\varphi = \operatorname{Re} f$ in D, then f and $\operatorname{Log} z$ have the same real part in the cut-plane $\mathbb{C} \setminus \{z : z + |z| = 0\}$ and so differ by a constant there; $f(z) = \operatorname{Log} z + a$ for some constant $a \in \mathbb{C}$ for all $z \in \mathbb{C} \setminus \{z : z + |z| = 0\}$. But this is not possible because the left hand side has a limit on the negative real axis (with 0 excluded) whereas the right hand side does not.

15.2 Local Existence of a Harmonic Conjugate

Notice that the domain in the example above is not star-like, and it is the presence of holes which allows for such a counterexample. For star-like domains, every harmonic function is indeed the real part of some suitable analytic function, as we now show.

Definition 15.2   Let φ be harmonic in a domain D. A function $\psi(x, y)$ is said to be a harmonic conjugate for φ in D if the function $f(x + iy) = \varphi(x, y) + i\psi(x, y)$ is analytic in D.

The analyticity of f implies that ψ is also harmonic in D. Furthermore, if $w(x, y)$ is also a harmonic conjugate for φ in D, then the functions $\varphi + i\psi$ and $\varphi + iw$ are analytic in D and have the same real part and so they differ by a constant. In other words, if φ possesses a harmonic conjugate in D, then it is unique to within a constant.

We can now prove the main result here.

Theorem 15.1   *Suppose that u is harmonic in a star-domain D. Then u possesses a harmonic conjugate there, that is, there is some f analytic in D such that $u(x, y) = \operatorname{Re} f(z)$ for all $z = x + iy \in D$.*

Proof.   For any $z = x + iy \in D$, set $g(z) = u_x(x, y) - iu_y(x, y)$. Now, since u is harmonic in D, it follows that

$$(u_x)_x = u_{xx} = -u_{yy} = (-u_y)_y \text{ and } (u_x)_y = (u_y)_x = -(-u_y)_x$$

so the real and imaginary parts of g satisfy the Cauchy-Riemann equations in D. Furthermore, all partial derivatives are continuous in D (because u is harmonic) and so we deduce that g is analytic in D.

By hypothesis, D is a star-domain and so g possesses a primitive in D, that is, there is $G \in H(D)$ such that $G' = g$ in D.

Let $G = U + iV$ and set $f = u + iV$. We shall show that $f \in H(D)$. Indeed, general theory tells us that

$$G' = U_x + iV_x = V_y - iU_y$$

but we know that

$$G' = g = u_x - iu_y$$

so that $u_x = V_y$ and $u_y = -V_x$. In other words, the real and imaginary parts of f obey the Cauchy-Riemann equations in D. Moreover, the partial derivatives of u and V are continuous in D (because u is harmonic and G is analytic) and so $f \in H(D)$ and the proof is complete. $\square$

Corollary 15.1 *Let u be harmonic in a domain D. For any $z_0 \in D$, there is some $r > 0$ such that u has a harmonic conjugate in the disc $D(z_0, r)$.*

Proof. For given $z_0 \in D$, there is some $r > 0$ such that $D(z_0, r) \subseteq D$. Since $D(z_0, r)$ is star-like, u has a harmonic conjugate there. $\square$

15.3 Maximum and Minimum Principle

Corollary 15.2 *Suppose that u is harmonic and is non-constant in a domain D. Then u has neither local maxima nor local minima in D.*

Proof. Suppose that u does have a local maximum at some point, say at $z_0 = x_0 + iy_0 \in D$. Then there is $r > 0$ and $f \in H(D(z_0, r))$ such that $u = \operatorname{Re} f$ in $D(z_0, r)$. Let $g = e^f$. Then $g \in H(D(z_0, r))$. Moreover, for any given $z = x + iy \in D(z_0, r)$,

$$|g(z)| = e^{u(x,y)} \le e^{u(x_0,y_0)} = |g(z_0)|$$

and so g is constant in $D(z_0, r)$ by the Maximum Modulus Principle. In particular, $e^{f(z)} = g(z) = g(z_0) = e^{f(z_0)}$ and so $(f(z) - f(z_0))/2\pi i \in \mathbb{Z}$ for all $z \in D(z_0, r)$. Since f is continuous, this means that f is also constant in the disc $D(z_0, r)$ and so f is constant on D, by the Identity Theorem. But then $u = \operatorname{Re} f$ is constant on D, a contradiction. We conclude that u has no local maxima.

By replacing u by $-u$ in the above, we see that u also has no local minima. $\square$

Corollary 15.3   *Suppose that u is harmonic and non-constant in $\mathbb{R}^2$. Then u is neither bounded from above nor from below.*

Proof.   First we note that, by the theorem, there is some function f analytic in $\mathbb{C}$ such that $u(x,y) = \operatorname{Re} f(x + iy)$ for all $(x,y) \in \mathbb{R}^2$. Now, suppose there is some constant M such that $u(x,y) \leq M$ for all $(x,y) \in \mathbb{R}^2$. For $z = x + iy \in \mathbb{C}$, set $g(z) = \exp(f(z))$. Then $|g(x + iy)| = e^{u(x,y)} \leq e^M$ and so g is a bounded entire function. By Liouville's Theorem, g and therefore f is constant. But then so is u and we have a contradiction. It follows that u is not bounded from above.

Applying the above argument to $-u$, we see that $-u$ is also not bounded from above, that is, u is not bounded from below.   □

Chapter 16

Local Properties of Analytic Functions

16.1 Local Uniform Convergence

We begin with a definition.

Definition 16.1 A sequence of functions (f_n) converges uniformly to f on the set A if and only if for any $\varepsilon > 0$ there is $N \in \mathbb{N}$ such that $n > N$ implies that $|f_n(z) - f(z)| < \varepsilon$ for any $z \in A$. (The important point is that the same N works no matter which $z \in A$ is selected.)

For any given domain D, we say that the sequence (f_n) of complex-valued functions on D converges locally uniformly to the function f on D if and only if for each point $z_0 \in D$ there is some $r > 0$ such that $D(z_0, r) \subseteq D$ and (f_n) converges uniformly to f on $D(z_0, r)$.

We show next that local uniform convergence is equivalent to uniform convergence on compact sets.

Theorem 16.1 *The sequence (f_n) converges locally uniformly to f on D if and only if (f_n) converges uniformly to f on any compact set $K \subseteq D$.*

Proof. Suppose that the sequence (f_n) converges to f uniformly on any compact set K in D and let $z_0 \in D$. Then there is some $R > 0$ such that $D(z_0, R) \subseteq D$. In particular, $\overline{D(z_0, R/2)} \subset D$ and (f_n) converges uniformly to f on the compact set $\overline{D(z_0, R/2)}$. In particular, (f_n) converges uniformly to f on the disc $D(z_0, R/2)$ and so it follows that (f_n) converges locally uniformly to f on D.

Conversely, suppose that (f_n) converges to f locally uniformly on D and let K be a given compact subset of D. For each point $z \in K$, there is some $r_z > 0$ such that $D(z, r_z) \subseteq D$ and such that $f_n \to f$ uniformly on the disc $D(z, r_z)$. The collection $\{ D(z, r_z) : z \in K \}$ is an open cover of K

and so has a finite subcover;

$$K \subseteq D(z_1, r_{z_1}) \cup \cdots \cup D(z_m, r_{z_m})$$

for some $z_1, \ldots, z_m \in K$. Let $\varepsilon > 0$ be given. Then for each $j = 1, \ldots, m$, there is some $N_j \in \mathbb{N}$ such that $n > N_j$ implies that

$$|f_n(z) - f(z)| < \varepsilon$$

for any $z \in D(z_j, r_{z_j})$. Setting $N = \max\{ N_j : 1 \leq j \leq m \}$, we see that $n > N$ implies that

$$|f_n(z) - f(z)| < \varepsilon$$

for any $z \in K$. That is, (f_n) converges to f uniformly on K.

An alternative proof of this, without using open covers, can be given as follows. Suppose that $f_n \to f$ locally uniformly on D but that (f_n) does not converge uniformly on some compact subset K of D. Then there is some $\varepsilon_0 > 0$ such that for any $N \in \mathbb{N}$ there is some $n > N$ and some point, ζ, say, in K (and depending on N) such that

$$|f_n(\zeta) - f(\zeta)| \geq \varepsilon_0.$$

We construct a sequence (w_j) in K as follows. For $N = 1$, there is $n_1 > 1$ and some z_{n_1}, say, in K such that

$$|f_{n_1}(z_{n_1}) - f(z_{n_1})| \geq \varepsilon_0.$$

Let $w_1 = z_{n_1}$. Then setting $N = n_1$, we may say that there is some $n_2 > n_1$ and some $z_{n_2} \in K$ such that

$$|f_{n_2}(z_{n_2}) - f(z_{n_2})| \geq \varepsilon_0.$$

Let $w_2 = z_{n_2}$. Now let $N = n_2$. Then there is some $n_3 > n_2$ and $z_{n_3} \in K$ such that

$$|f_{n_3}(z_{n_3}) - f(z_{n_3})| \geq \varepsilon_0.$$

Let $w_3 = z_{n_3}$. Continuing in this way, we obtain a sequence (w_j) in K (and integers $n_1 < n_2 < \ldots$ in $\mathbb{N}$) such that

$$\left| f_{n_j}(w_j) - f(w_j) \right| \geq \varepsilon_0 \qquad (*)$$

for all $j \in \mathbb{N}$.

Since the sequence (w_j) lies in K and K is compact, there is a convergent subsequence, $w_{j_k} \to \zeta$, say, in K as $k \to \infty$.

But since $f_n \to f$ locally uniformly, there is some $r > 0$ such that $f_n \to f$ uniformly in $D(\zeta, r)$. In particular, there is $k_0 \in \mathbb{N}$ such that if $k > k_0$ then $|f_{n_{j_k}}(z) - f(z)| < \varepsilon_0$ for all $z \in D(\zeta, r)$. However, for sufficiently large k, $w_{j_k} \in D(\zeta, r)$ and so

$$\left|f_{n_{j_k}}(w_{j_k}) - f(w_{j_k})\right| < \varepsilon_0$$

for sufficiently large k. This contradicts (*) and we conclude that $f_n \to f$ uniformly on K, as required. $\qquad \square$

Proposition 16.1 *Suppose that $f_n \to f$ locally uniformly on a domain D and that each f_n is continuous. Then $f : D \to \mathbb{C}$ is continuous.*

Proof. For any given $z_0 \in D$ there is some $R > 0$ such that $f_n \to f$ uniformly in the disc $D(z_0, R)$. Let $\varepsilon > 0$ be given. Then there is some N such that $|f_N(\zeta) - f(\zeta)| < \frac{1}{3}\varepsilon$ for all $\zeta \in D(z_0, R)$. Since f_N is continuous, there is $\rho > 0$ such that $|f(w) - f(z_0)| < \frac{1}{3}\varepsilon$ whenever $w \in D(z_0, \rho)$. Set $r = \min\{R, \rho\}$. Then for any $w \in D(z_0, r)$, we have

$$
\begin{aligned}
|f(w) - f(z_0)| &\leq |f(w) - f_N(w)| + |f_N(w) - f_N(z_0)| + |f_N(z_0) - f(z_0)| \\
&< \tfrac{1}{3}\varepsilon + \tfrac{1}{3}\varepsilon + \tfrac{1}{3}\varepsilon \\
&= \varepsilon.
\end{aligned}
$$

It follows that f is continuous at z_0 and so the proof is complete. $\qquad \square$

Proposition 16.2 *Suppose that (f_n) and (g_n) are sequences of continuous functions such that $f_n \to f$ and $g_n \to g$ locally uniformly on the domain D. Then*

(i) $\alpha f_n \to \alpha f$ *locally uniformly on D, for any $\alpha \in \mathbb{C}$.*
(ii) $f_n + g_n \to f + g$ *locally uniformly on D.*
(iii) $f_n g_n \to fg$ *locally uniformly on D.*
(iv) *If, furthermore, each $g_n \neq 0$ and $g \neq 0$ on D, then $\frac{f_n}{g_n} \to \frac{f}{g}$ locally uniformly on D.*

Proof. It is enough to show uniform convergence on any given compact set K in D and it is straightforward to show that this is true of αf_n to αf and $f_n + g_n$ to $f + g$.

Let us show that $f_n g_n \to fg$ uniformly on K. The uniform convergence on K implies that both f and g are continuous on K and therefore bounded there. Hence there is $M > 0$ such that $|f(z)| \leq M$ and $|g(z)| \leq M$, for all

$z \in K$. Also, the uniform convergence of f_n to f on K implies that

$$|f_n(z)| \leq |f_n(z) - f(z)| + |f(z)|$$
$$\leq 1 + M$$

for all sufficiently large n and for any $z \in K$. But then, for given $\varepsilon > 0$,

$$|f_n(z)g_n(z) - f(z)g(z)| \leq |f_n(z)| \, |g_n(z) - g(z)| + |f_n(z) - f(z)| \, |g(z)|$$
$$\leq (M+1)\varepsilon + M\varepsilon$$

for all sufficiently large n and any $z \in K$. It follows that $f_n g_n \to fg$ uniformly on K and hence locally uniformly on D.

To show that $f_n/g_n \to f/g$ locally uniformly, we need only show that $1/g_n \to 1/g$ locally uniformly (and then apply the argument above). So suppose that $g_n \neq 0$ and $g \neq 0$ on D. Let $K \subset D$ be compact. As above, we know that g is continuous on K. It follows that $|g|$ attains its lower bound on K and so there is $m > 0$ such that $|g(z)| > m$, for all $z \in K$.

Furthermore, $|g(z) - g_n(z)| < \frac{1}{2}m$ for sufficiently large n and all $z \in K$ and so

$$m \leq |g(z)| \leq |g(z) - g_n(z)| + |g_n(z)| < \tfrac{m}{2} + |g_n(z)|,$$

giving $|g_n(z)| > \frac{1}{2}m$ for large n and any $z \in K$. Hence

$$\left| \frac{1}{g_n(z)} - \frac{1}{g(z)} \right| = \left| \frac{g(z) - g_n(z)}{g_n(z)\,g(z)} \right| \leq \frac{2\,|g(z) - g_n(z)|}{m^2},$$

for sufficiently large n and any $z \in K$. The result now follows because $g_n \to g$ uniformly on K. $\qquad \square$

16.2 Hurwitz's Theorem

Locally uniformly convergent sequences of analytic functions are very well-behaved, as we see next.

Theorem 16.2 *Suppose that D is a domain and that (f_n) is a sequence in $H(D)$ which converges locally uniformly to f on D. Then $f \in H(D)$ and moreover, for each $k \in \mathbb{N}$, the sequence of derivatives $(f_n^{(k)})$ converges locally uniformly to the kth derivative $f^{(k)}$ on D.*

Proof. First we shall show that f is analytic in D. Let $z_0 \in D$. Then there is $r > 0$ such that $D(z_0, r) \subseteq D$. Let T be any triangle in $D(z_0, r)$.

By Cauchy's Theorem,

$$\int_{\partial T} f_n = 0$$

for every $n \in \mathbb{N}$. Now, (f_n) converges to f uniformly on the compact set ∂T and so f is continuous on ∂T. Let $\varepsilon > 0$ be given. Then there is N such that $|f_n(z) - f(z)| < \varepsilon$ for any $z \in \partial T$, whenever $n > N$. Hence

$$\left| \int_{\partial T} f - \int_{\partial T} f_n \right| = \left| \int_{\partial T} (f - f_n) \right|$$
$$\leq \varepsilon \, L(\partial T)$$

by the Basic Estimate, theorem 7.1. It follows that

$$\int_{\partial T} f = \lim_{n \to \infty} \int_{\partial T} f_n = 0.$$

By Morera's Theorem, theorem 8.8, we conclude that f is analytic on D.

To show that the derivatives of f_n converge to those of the function f locally uniformly, we use Cauchy's Integral Formulae. Let $z_0 \in D$ and suppose $R > 0$ is such that $D(z_0, 2R) \subseteq D$. We will show that $f_n^{(k)} \to f^{(k)}$ uniformly on $D(z_0, R)$.

Let $k \in \mathbb{N}$ and $z \in D(z_0, R)$ and let $\varepsilon > 0$. Then, by Cauchy's Integral Formula

$$| f^{(k)}(z) - f_n^{(k)}(z) | = \left| \frac{1}{2\pi i} \int_\gamma \frac{f(w) - f_n(w)}{(w - z)^{k+1}} \, dw \right|$$

where we take γ to be the circle $\gamma(t) = z_0 + \frac{3}{2} R e^{2\pi i t}$, $0 \leq t \leq 1$. Now, for all sufficiently large n, $|f(w) - f_n(w)| < \varepsilon$ for all w in the compact set tr γ. Moreover, for all such w, $|w - z| \geq \frac{1}{2} R$. Hence, by the Basic Estimate, theorem 7.1, we see that the right hand side above is bounded by

$$\left| \frac{1}{2\pi i} \int_\gamma \frac{f(w) - f_n(w)}{(w - z)^{k+1}} \, dw \right| \leq \frac{1}{2\pi} \frac{\varepsilon}{(\frac{1}{2} R)^{k+1}} \, 2\pi \frac{3}{2} R,$$

provided n is sufficiently large. The result follows. □

Theorem 16.3 (Hurwitz's Theorem) *Suppose that (f_n) is a sequence of functions analytic on the domain D such that $f_n \to f$ locally uniformly on D. Suppose, further, that $f_n \neq 0$ in D. Then either f is identically zero on D, or $f \neq 0$ on D.*

Proof.   We know that $f \in H(D)$. Suppose that f is not identically zero in D. Let $z_0 \in D$. Then there is $\rho > 0$ such that $D(z_0, \rho) \subseteq D$. Since f is not identically zero, there is $0 < r < \rho$ such that $f(z) \neq 0$ for all z with $|z - z_0| = r$. (Otherwise, z_0 would be a limit point of zeros of f.)

Since the circle $C = \{\, z : |z - z_0| = r \,\}$ is compact and f is continuous, there is $\zeta \in C$ such that $|f(z)| \geq |f(\zeta)|$ for all $z \in C$. By construction, f is not zero on C and so $m = |f(\zeta)| > 0$, i.e., $|f(z)| \geq m > 0$ on C.

Let n be so large that $|f(z) - f_n(z)| < m$ for all $z \in C$. (This is possible, since $f_n \to f$ uniformly on compact sets.) Then

$$|f(z)| \geq m > |f(z) - f_n(z)|$$

on the circle C. By Rouché's Theorem, theorem 12.4, it follows that f and f_n have the same number of zeros inside the circle C, that is, none. In particular, $f(z_0) \neq 0$ which is to say that f has no zeros in D.   □

Theorem 16.4   *Suppose $f_n \in H(D)$ and that $f_n \to f$ locally uniformly on the domain D. Suppose that $z_0 \in D$ is a zero of f of order m, then for any $R > 0$ there is a disc $D(z_0, r)$, with $r < R$, and $N \in \mathbb{N}$ such that for all $n > N$, f_n has exactly m zeros in $D(z_0, r)$.*

Proof.   Let $R > 0$ be given. Since $f_n \to f$ locally uniformly, there is some disc $D(z_0, \rho)$ such that $f_n \to f$ uniformly on $D(z_0, \rho)$. Then, arguing as above, we deduce that there is $0 < r < \min\{\rho, R\}$ and $\alpha > 0$ such that $|f(z)| \geq \alpha$ for all $z \in C = \{|z - z_0| = r\}$. Moreover, we may assume that z_0 is the only zero of f inside the circle C (because the zeros of f are isolated).

The uniform convergence on compact sets implies that there is $N \in \mathbb{N}$ such that if $n > N$ then

$$|f(z) - f_n(z)| < \alpha$$

for all $z \in C$. Hence

$$|f| \geq \alpha > |f - f_n|$$

on C and so, by Rouché's Theorem, f and f_n have the same number of zeros inside the circle C, namely m.   □

16.3   Vitali's Theorem

Definition 16.2   A sequence (f_n) of functions is said to be locally uniformly bounded on a domain D if for each $z_0 \in D$ there is some $r > 0$ with $D(z_0, r) \subseteq D$ and some $M > 0$ such that $|f_n(z)| < M$ for all n and all $z \in D(z_0, r)$. Note that both r and M may depend on z_0.

The result of interest in this connection is Vitali's Theorem (which we will not prove here), as follows.

Theorem 16.5   (Vitali's Theorem)   *Let (f_n) be a sequence of functions analytic in a domain D. Suppose that (f_n) is locally uniformly bounded in D and that there is some set A in D such that A has a limit point in D and such that $(f_n(z))$ converges for all $z \in A$. Then there is a function f such that $f_n \to f$ locally uniformly in D (so that, in particular, $f \in H(D)$).*

Appendix A

Some Results from Real Analysis

We collect here some of the basic results from real analysis that we have needed. They all depend crucially on the Completeness Property of $\mathbb{R}$. We begin with some definitions.

A.1 Completeness of $\mathbb{R}$

Definition A.3 A non-empty subset S of $\mathbb{R}$ is said to be bounded from above if there is some $M \in \mathbb{R}$ such that $a \leq M$ for all $a \in S$. Any such number M is called an upper bound for the set S.

The non-empty subset S of $\mathbb{R}$ is said to be bounded from below if there is some $m \in \mathbb{R}$ such that $m \leq a$ for all $a \in S$. Any such number m is called a lower bound for the set S.

A subset of $\mathbb{R}$ is said to be bounded if it is bounded both from above and from below.

Suppose S is a non-empty subset of $\mathbb{R}$ which is bounded from above. The number M is the least upper bound of S (lub S) if

(i) $a \leq M$ for all $a \in S$ (i.e., M is an upper bound for S).
(ii) If M' is any upper bound for S, then $M \leq M'$.

If S is a non-empty subset of $\mathbb{R}$ which is bounded from below, then the number m is the greatest lower bound of S (glb S) if

(i) $m \leq a$ for all $a \in S$ (i.e., m is a lower bound for S).
(ii) If m' is any lower bound for S, then $m' \leq m$.

Note that the least upper bound and the greatest lower bound of a set S need not themselves belong to S. They may or they may not. The least upper bound is also called the supremum (sup) and the greatest lower bound is also called the infimum (inf).

Evidently, if M is an upper bound for S, then so is any number greater than M. It is also clear that M is an upper bound for any non-empty subset of S. In particular, $\sup S$ is an upper bound for any such subset of S. Note that if $M = \operatorname{lub} S$, then there is some sequence (x_n) in S such that $x_n \to M$, as $n \to \infty$. (Indeed, for any $n \in \mathbb{N}$, the number $M - \frac{1}{n}$ fails to be an upper bound for S and so there is some $x_n \in S$ such that $x_n > M - \frac{1}{n}$. Hence x_n obeys $M - \frac{1}{n} < x_n \le M$ which demands $x_n \to M$.) Analogous remarks apply to lower bounds and $\operatorname{glb} S$.

Example A.1 $\sup[0,1] = 1 = \sup(0,1)$ and $\inf[0,1] = 0 = \inf(0,1)$. Note that $(0,1)$ has neither a maximum element nor a minimum element.

The essential property which distinguishes $\mathbb{R}$ from $\mathbb{Q}$ is the following.

The Completeness Property of $\mathbb{R}$
Any non-empty subset of $\mathbb{R}$ which is bounded from above possesses a least upper bound.

A consequence of this property, for example, is that any positive real number possesses a square root. In particular, thanks to this we can be confident that $\sqrt{2}$ exists as a real number. (It is given by $\sup\{\, x : x^2 < 2 \,\}$.)

Proposition A.3   *If (a_n) is an increasing sequence of real numbers and is bounded from above, then it converges.*

Proof.   By hypothesis, $\{\, a_n : n \in \mathbb{N} \,\}$ is bounded from above. Let $K = \operatorname{lub}\{\, a_n : n \in \mathbb{N} \,\}$. We claim that $a_n \to K$ as $n \to \infty$.

Let $\varepsilon > 0$ be given. Since K is *an* upper bound for $\{\, a_n : n \in \mathbb{N} \,\}$, it follows that $a_n \le K$ for all n. On the other hand, $K - \varepsilon < K$ and K is the least upper bound of $\{\, a_n : n \in \mathbb{N} \,\}$ and so $K - \varepsilon$ is not an upper bound for $\{\, a_n : n \in \mathbb{N} \,\}$. This means that there is some a_j, say, with $a_j > K - \varepsilon$. But the sequence (a_n) is increasing and so $a_n \ge a_j$ for all $n > j$. Hence $a_n > K - \varepsilon$ for all $n > j$. We have shown that

$$ K - \varepsilon < a_n \le K < K + \varepsilon $$

for all $n > j$ and so the proof is complete. $\square$

Corollary A.1   *Any sequence (b_n) in $\mathbb{R}$ which is decreasing and bounded from below must converge.*

Proof. Just apply the above result to the sequence $a_n = -b_n$. □

In fact, b_n converges to the greatest lower bound of $\{\, b_n : n \in \mathbb{N} \,\}$.

A.2 Bolzano-Weierstrass Theorem

Theorem A.6 (Bolzano-Weierstrass Theorem) *Any bounded sequence of real numbers possesses a convergent subsequence.*

Proof. Suppose that M and m are upper and lower bounds for (a_n),

$$m \le a_n \le M .$$

We construct a certain bounded decreasing sequence and use the fact that this converges to its greatest lower bound and so drags a suitable subsequence of (a_n) along with it.

To construct the first element of the auxiliary decreasing sequence, let $M_1 = \mathrm{lub}\{\, a_n : n \in \mathbb{N} \,\}$. Then $M_1 - 1$ is *not* an upper bound for $\{\, a_n : n \in \mathbb{N} \,\}$ and so there must be some n_1, say, in $\mathbb{N}$ such that

$$M_1 - 1 < a_{n_1} \le M_1 .$$

Next, we construct M_2 as follows. Let $M_2 = \mathrm{lub}\{\, a_n : n > n_1 \,\}$ so that $M_2 \le M_1$. Moreover, $M_2 - \frac{1}{2}$ is *not* an upper bound for $\{\, a_n : n > n_1 \,\}$ and so there is some $n_2 > n_1$ such that

$$M_2 - \tfrac{1}{2} < a_{n_2} \le M_2 .$$

Continuing in this way, we construct a sequence $(M_j)_{j \in \mathbb{N}}$ and a sequence $(n_j)_{j \in \mathbb{N}}$ such that $M_{j+1} \le M_j$, $n_{j+1} > n_j$, and

$$M_j - \tfrac{1}{j} < a_{n_j} \le M_j$$

for all $j \in \mathbb{N}$.

Now, $m \le a_{n_j} \le M_j$ and so (M_j) is a decreasing sequence which is bounded from below. It follows that (M_j) converges, say $M_j \to \mu$, as $j \to \infty$. However, by our very construction,

$$M_j - \tfrac{1}{j} < a_{n_j} \le M_j$$

and so $a_{n_j} \to \mu$, as $j \to \infty$, and the proof is complete. □

Remark A.1   Note that if $a_n \in [a, b]$ for all n, then the limit of any convergent subsequence also belongs to the interval $[a, b]$.

Recall that a sequence (a_n) in $\mathbb{R}$ is a Cauchy sequence if for any given $\varepsilon > 0$ there is $N \in \mathbb{N}$ such that $|a_n - a_m| < \varepsilon$ whenever both $n, m > N$.

Theorem A.7   *Every Cauchy sequence in $\mathbb{R}$ converges.*

Proof.   First, we show that any Cauchy sequence (a_n) in $\mathbb{R}$ must be bounded. Indeed, we know that there is some $N \in \mathbb{N}$ such that both $n > N$ and $m > N$ imply that

$$|a_n - a_m| < 1.$$

In particular, for any $j > N$,

$$|a_j| \le |a_j - a_{N+1}| + |a_{N+1}| < 1 + |a_{N+1}|.$$

It follows that if $M = 1 + \max\{\, |a_1|, |a_2|, \ldots, |a_{N+1}| \,\}$, then

$$|a_k| \le M$$

for all $k \in \mathbb{N}$, i.e., (a_n) is bounded.

To show that (a_n) converges, we note that by the Bolzano-Weierstrass Theorem, (a_n) has some convergent subsequence, say $a_{n_k} \to \alpha$, as $k \to \infty$. We show that $a_n \to \alpha$.

Let $\varepsilon > 0$ be given. Then there is $k_0 \in \mathbb{N}$ such that $k > k_0$ implies that

$$|a_{n_k} - \alpha| < \tfrac{1}{2}\varepsilon.$$

Since (a_n) is a Cauchy sequence, there is N_0 such that both $n > N_0$ and $m > N_0$ imply that

$$|a_n - a_m| < \tfrac{1}{2}\varepsilon.$$

Let $N = \max\{\, k_0, N_0 \,\}$. Then

$$|a_n - \alpha| \le |a_n - a_{n_k}| + |a_{n_k} - \alpha| < \tfrac{1}{2}\varepsilon + \tfrac{1}{2}\varepsilon = \varepsilon$$

whenever $n > N$. Thus $a_n \to \alpha$ as $k \to \infty$ as required. $\square$

Remark A.2   Note that a convergent sequence is necessarily a Cauchy sequence. Indeed, if $a_n \to \alpha$, then the inequality

$$|a_n - a_m| \le |a_n - \alpha| + |\alpha - a_m|$$

shows that (a_n) is a Cauchy sequence.

A.3 Comparison Test for Convergence of Series

The above result enables us to conclude that various series converge even though we may not know their sum.

Theorem A.8 (Comparison Test) *Suppose that $a_0, a_1, \ldots$ and $b_0, b_1, \ldots$ are sequences in $\mathbb{R}$ such that $0 \le a_n \le b_n$ for all $n = 0, 1, 2, \ldots$. If the series $\sum_{n=0}^{\infty} b_n$ converges, then so does the series $\sum_{n=0}^{\infty} a_n$.*

Proof. Let $S_n = \sum_{k=0}^{n} a_k$ and $T_n = \sum_{k=0}^{n} b_k$ be the partial sums. Then we see that for $n > m$

$$0 \le S_n - S_m = \sum_{k=m+1}^{n} a_k \le \sum_{k=m+1}^{n} b_k = T_n - T_m.$$

If $\sum_{n=0}^{\infty} b_n$ is convergent, then (T_n) is a Cauchy sequence and so therefore is (S_n). Hence $\sum_{n=0}^{\infty} a_n$ converges. □

A.4 Dirichlet's Test

Theorem A.9 (Dirichlet's Test) *Suppose $a_0, a_1, \ldots$ is a sequence in $\mathbb{R}$ such that the partial sums $S_n = \sum_{k=0}^{n} a_k$ are bounded and suppose $y_0 \ge y_1 \ge y_2 \ge \cdots \ge 0$ is a decreasing positive sequence in $\mathbb{R}$ such that $y_n \downarrow 0$, as $n \to \infty$. Then $\sum_{n=0}^{\infty} a_n y_n$ is convergent.*

Proof. The proof uses a rearrangement trick (called "summation by parts"). Let $T_n = \sum_{k=0}^{n} a_k y_k$. Then, for $n > m$ (by straightforward verification),

$$T_n - T_m = \sum_{k=m+1}^{n} a_k y_k = S_n y_{n+1} - S_m y_{m+1} + \sum_{k=m+1}^{n} S_k (y_k - y_{k+1}).$$

Let $\varepsilon > 0$ be given. By hypothesis, there is $M > 0$ such that $|S_n| \le M$, for all n and since $y_n \downarrow 0$, as $n \to \infty$, there is $N \in \mathbb{N}$ such that $0 \le y_n < \varepsilon/2M$, for all $n > N$. Hence, if $n, m > N$,

$$|T_n - T_m| \leq |S_n y_{n+1}| + |S_m y_{m+1}| + \sum_{k=m+1}^{n} |S_k| (y_k - y_{k+1})$$

$$\leq M y_{n+1} + M y_{m+1} + \sum_{k=m+1}^{n} M(y_k - y_{k+1})$$

$$= M y_{n+1} + M y_{m+1} + M(y_{m+1} - y_{n+1})$$

$$= 2M y_{m+1}$$

$$< \varepsilon.$$

It follows that (T_n) is a Cauchy sequence in $\mathbb{R}$ and therefore $\sum_{n=0}^{\infty} a_n y_n$ is convergent. $\qquad\square$

A.5   Alternating Series Test

The Alternating Series Test is an immediate consequence, as follows.

Theorem A.10 (Alternating Series Test)   *Suppose* $x_0 \geq x_1 \geq x_2 \geq \cdots \geq 0$ *is a decreasing positive sequence in* $\mathbb{R}$ *such that* $x_n \downarrow 0$ *as* $n \to \infty$. *Then the alternating series* $x_0 - x_1 + x_2 - x_3 + \cdots = \sum_{n=0}^{\infty} (-1)^n x_n$ *converges.*

Proof.   Set $a_n = (-1)^n$ and $y_n = x_n$. Then the partial sums $S_n = \sum_{k=0}^{n} a_n = \frac{1}{2}(1 + (-1)^n)$ are bounded so we can apply Dirichlet's Test to obtain convergence, as stated. $\qquad\square$

A.6   Continuous Functions on $[a, b]$ Attain their Bounds

Next, we look at properties of continuous functions.

Theorem A.11   *Suppose that the function* $f : [a, b] \to \mathbb{R}$ *is continuous. Then* f *is bounded on* $[a, b]$ *and achieves its maximum and minimum on* $[a, b]$.

Proof.   We argue by contradiction. Suppose that f is continuous on $[a, b]$ but is not bounded from above. This means that for any given M whatsoever, there will be some $x \in [a, b]$ such that $f(x) > M$. In particular, for each $n \in \mathbb{N}$ there is some a_n, say, with $a_n \in [a, b]$ such that $f(a_n) > n$.

   The sequence $(a_n)_{n \in \mathbb{N}}$ lies in the bounded interval $[a, b]$ and so, by the Bolzano-Weierstrass Theorem, it has a convergent subsequence $(a_{n_k})_{k \in \mathbb{N}}$, say; $a_{n_k} \to \alpha$ as $k \to \infty$. Since $a \leq a_{n_k} \leq b$ for all k, it follows that $a \leq \alpha \leq b$. By hypothesis, f is continuous at α and so $a_{n_k} \to \alpha$ implies that $f(a_{n_k}) \to f(\alpha)$. But any convergent sequence is bounded (mimic the argument above for Cauchy sequences). This is a contradiction, so we conclude that f must be bounded from above.

To show that f is also bounded from below, consider $g = -f$. Then g is continuous because f is. The argument just presented, applied to g, shows that g is bounded from above. But this just means that f is bounded from below, as required.

We must now show that there is $\alpha, \beta \in [a, b]$ such that $f(x) \leq f(\alpha)$ and $f(x) \geq f(\beta)$ for all $x \in [a, b]$.

We know that f is bounded so let $M = \sup\{ f(x) : x \in [a, b] \}$. Then $f(x) \leq M$ for all $x \in [a, b]$. We show that there is some $\alpha \in [a, b]$ such that $f(\alpha) = M$. To see this, suppose there is no such α. Then $f(x) < M$ and so, in particular, $M - f$ is continuous and strictly positive on $[a, b]$. It follows that $h = 1/(M - f)$ is continuous and positive on $[a, b]$ and so is bounded, by the first part. Therefore there is some constant K such that $0 < h \leq K$ on $[a, b]$, that is,

$$0 < \frac{1}{M - f} \leq K.$$

Hence $f \leq M - 1/K$ which says that $M - 1/K$ is an upper bound for f on $[a, b]$. But then this contradicts the fact that M is the least upper bound of f on $[a, b]$. We conclude that f must achieve this bound, i.e., there is $\alpha \in [a, b]$ such that $f(\alpha) = M = \sup\{ f(x) : x \in [a, b] \}$.

In a similar way, if f does not achieve its greatest lower bound, m, then $f - m$ is continuous and strictly positive on $[a, b]$. Hence there is L such that

$$0 < \frac{1}{f - m} \leq L$$

on $[a, b]$. Hence $m + 1/L \leq f$ and $m + 1/L$ is a lower bound for f on $[a, b]$. This contradicts the fact that m is the greatest lower bound for f on $[a, b]$ and we can conclude that f does achieve its greatest lower bound, that is, there is $\beta \in [a, b]$ such that $f(\beta) = m$. □

Remark A.3 Note that neither α nor β need be unique.

A.7   Intermediate Value Theorem

Theorem A.12 (Intermediate Value Theorem)   *Any real-valued function f continuous on the interval $[a, b]$ assumes all values between $f(a)$ and $f(b)$. In other words, if ζ lies between the values $f(a)$ and $f(b)$, then there is some s with $a \leq s \leq b$ such that $f(s) = \zeta$.*

Proof.   Suppose f is continuous on $[a, b]$ and let ζ be any value between $f(a)$ and $f(b)$. If $\zeta = f(a)$, take $s = a$ and if $\zeta = f(b)$ take $s = b$.

Suppose that $f(a) < f(b)$ and let $f(a) < \zeta < f(b)$. Let A be the set $A = \{ x \in [a, b] : f(x) < \zeta \}$. Then $a \in A$ and so A is a non-empty subset of the bounded interval $[a, b]$. Hence A is bounded and so has a least upper bound, s, say. We shall show that $f(s) = \zeta$.

Since $s = \operatorname{lub} A$, there is some sequence (a_n) in A such that $a_n \to s$. But $A \subseteq [a, b]$ and so $a \leq a_n \leq b$ and it follows that $a \leq s \leq b$. Furthermore, by the continuity of f at s, it follows that $f(a_n) \to f(s)$. However, $a_n \in A$ and so $f(a_n) < \zeta$ for each n and it follows that $f(s) \leq \zeta$. Since, in addition, $\zeta < f(b)$, we see that $s \neq b$ and so we must have $a \leq s < b$.

Let (t_n) be any sequence in (s, b) such that $t_n \to s$. Since $t_n \in [a, b]$ and $t_n > s$, it must be the case that $t_n \notin A$, that is, $f(t_n) \geq \zeta$. Now, f is continuous at s and so $f(t_n) \to f(s)$ which implies that $f(s) \geq \zeta$. We deduce that $f(s) = \zeta$, as required.

Now suppose that $f(a) > \zeta > f(b)$. Set $g(x) = -f(x)$. Then we have that $g(a) < -\zeta < g(b)$ and applying the above result to g, we can say that there is $s \in [a, b]$ such that $g(s) = -\zeta$, that is $f(s) = \zeta$ and the proof is complete. $\square$

A.8   Rolle's Theorem

For differentiable functions, more can be said.

Theorem A.13 (Rolle's Theorem)   *Suppose that f is continuous on the closed interval $[a, b]$ and is differentiable in the open interval (a, b). Suppose further that $f(a) = f(b)$. Then there is some $\xi \in (a, b)$ such that $f'(\xi) = 0$. (Note that ξ need not be unique.)*

Proof.   Since f is continuous on $[a, b]$, it follows that f is bounded and attains its bounds, by Theorem A.11. Let $m = \inf\{ f(x) : x \in [a, b] \}$ and let $M = \sup\{\, f(x) : x \in [a,b]\,\}$, so that

$$m \leq f(x) \leq M\,, \quad \text{for all } x \in [a,b].$$

If $m = M$, then f is constant on $[a,b]$ and this means that $f'(x) = 0$ for *all* $x \in (a,b)$. In this case, any $\xi \in (a,b)$ will do.

Suppose now that $m \neq M$, so that $m < M$. Since $f(a) = f(b)$ at least one of m or M must be different from this common value $f(a) = f(b)$.

Suppose that $M \neq f(a)$ ($= f(b)$). As noted above, by theorem A.11, there is some $\xi \in [a,b]$ such that $f(\xi) = M$. Now, $M \neq f(a)$ and $M \neq f(b)$ and so $\xi \neq a$ and $\xi \neq b$. It follows that ξ belongs to the open interval (a,b).

We shall show that $f'(\xi) = 0$. To see this, we note that $f(x) \leq M = f(\xi)$ for any $x \in [a,b]$ and so (putting $x = \xi+h$) it follows that $f(\xi+h) - f(\xi) \leq 0$ provided $|h|$ is small enough to ensure that $\xi + h \in [a,b]$. Hence

$$\frac{f(\xi+h) - f(\xi)}{h} \leq 0 \quad \text{for } h > 0 \text{ and small} \qquad (*)$$

and

$$\frac{f(\xi+h) - f(\xi)}{h} \geq 0 \quad \text{for } h < 0 \text{ and small}. \qquad (**)$$

But $(*)$ approaches $f'(\xi)$ as $h \downarrow 0$ which implies that $f'(\xi) \leq 0$. On the other hand, $(**)$ approaches $f'(\xi)$ as $h \uparrow 0$ and so $f'(\xi) \geq 0$. Putting these two results together, we see that it must be the case that $f'(\xi) = 0$, as required.

It remains to consider the case when $M = f(a)$. This must require that $m < f(a)$ ($= f(b)$). We proceed now just as before to deduce that there is some $\xi \in (a,b)$ such that $f(\xi) = m$ and so $(*)$ and $(**)$ hold but *with the inequalities reversed*. However, the conclusion is the same, namely that $f'(\xi) = 0$. $\qquad\square$

A.9 Mean Value Theorem

Theorem A.14 (Mean Value Theorem) *Suppose that f is continuous on the closed interval $[a,b]$ and differentiable on the open interval (a,b). Then there is some $\xi \in (a,b)$ such that*

$$f'(\xi) = \frac{f(b) - f(a)}{b - a}\,.$$

Proof.   Let $y = \ell(x) = mx + c$ be the straight line passing through the pair of points $(a, f(a))$ and $(b, f(b))$. Then the slope m is equal to the ratio $(f(b) - f(a))/(b - a)$.

Let $g(x) = f(x) - \ell(x)$. Evidently, g is continuous on $[a, b]$ and differentiable on (a, b) (because ℓ is). Furthermore, since $\ell(a) = f(a)$ and $\ell(b) = f(b)$, by construction, we find that $g(a) = 0 = g(b)$. By Rolle's Theorem, theorem A.13, applied to g, there is some $\xi \in (a, b)$ such that $g'(\xi) = 0$. However, $g'(x) = f'(x) - m$ for any $x \in (a, b)$ and so

$$f'(\xi) = m = \frac{f(b) - f(a)}{b - a}$$

and the proof is complete.   □

We know that a function which is constant on an open interval is differentiable and that its derivative is zero. The converse is true.

Corollary A.2   *Suppose that f is differentiable on the open interval (a, b) and that $f'(x) = 0$ for all $x \in (a, b)$. Then f is constant on (a, b).*

Proof.   Let $\alpha < \beta$ be any pair of points in (a, b). Applying the Mean Value Theorem to f on $[\alpha, \beta]$, we may say that there is some $\xi \in (\alpha, \beta)$ such that

$$f'(\xi) = \frac{f(\beta) - f(\alpha)}{\beta - \alpha}.$$

However, f' vanishes on (a, b) and so $f'(\xi) = 0$ and therefore $f(\alpha) = f(\beta)$. The result follows.   □

Bibliography

The following list includes books that have been useful for the preparation of these notes as well as some which may be of interest for further study.

Ash, R. B. *Complex Variables.* New York, London, Academic Press, 1971.

Bak, J. and D. Newman. *Complex Analysis.* 2nd edition, Springer-Verlag, 1997.

Beardon, A. F. *Complex Analysis: the argument principle in analysis and topology.* Chichester, Wiley, 1979.

Conway, J. B. *Functions of One Complex Variable.* 2nd edition, Springer-Verlag, New York, 1978.

Duncan, J. *Elements of Complex Analysis.* London, J. Wiley, 1968.

Grove, E. and G. Ladas. *Introduction to Complex Variables.* Houghton Mifflin, Boston, 1974.

Hairer, E. and G. Wanner. *Analysis by Its History.* Springer-Verlag, New York 1996.

Marsden, J. E. and M. J. Hoffman. *Basic Complex Analysis.* 3rd edition, W. H. Freeman, 1999.

Rudin, W. *Real and Complex Analysis.* 3rd edition, New York, London, McGraw-Hill, 1987.

Stewart, I. and D. Tall. *Complex Analysis.* Cambridge University Press, 1993.

Tall, D. *Functions of a Complex Variable.* Routledge and Kegan Paul, London, Henley & Boston, 1977.

For real analysis, any of the following books make a very good start.

Abbott, S. *Understanding Analysis.* Springer-Verlag, 2001.

Bartle, R. G. and D. R. Sherbert. *Introduction to real Analysis.* J. Wiley & Sons Inc., 1982.

Rudin, W. *Principles of Mathematical Analysis.* International Series in Pure and Applied Mathematics, McGraw-Hill, New York, 1964.

For a history of the various characters in mathematics, see:

The MacTutor History of Mathematics archive, which is located at the url
http://www-groups.dcs.st-and.ac.uk/~history/
(created by John J. O'Connor and Edmund F. Robertson)

Index